Community, Change and Border Towns

This book provides an interdisciplinary approach to power, inclusion/exclusion and hierarchy in a Turkish border town, with a focus on the impact of nation-state border on social stratification and change.

Through the lens of ethnographic research and oral history, the book explores social mobility among various strata within the context of transition from Ottoman rule to the Republican regime, in order to reveal culturally informed strategies of border dwellers in coming to grips with new border contexts. It is suggested that the border perspective will move the social analysis beyond "methodological territorialism" and provide a theoretical framework that explores social change at the intersection of local, national and transnational processes.

This book will appeal to readers interested in borders and circulations, social structure and power relations in border regions, as well as transnational shadow networks in the Turkish/Middle Eastern context. The book is a valuable resource for students and scholars of border anthropology, political and economic geography, studies of globalization and transnationalism, anthropology of illegality and Turkish and Middle Eastern studies. It will be a useful grounding for humanitarian professionals who are learning about the social and economic landscape of border towns.

H. Pınar Şenoğuz is currently a post-doctoral fellow at the University of Göttingen in Germany. Until she was dismissed by the emergency decree in 2016, she taught sociology in a public university in Gaziantep city on the Turkish border with Syria. Among her published and forthcoming articles, she explores shadow markets and illegality, post-migration conflicts and refugee hospitality in the southeastern border regions of Turkey. Her research interests include the anthropology of borderland and illegality, border politics and refugee reception policies in EU and Middle Eastern countries, and power and in/exclusion in the Middle East.

Border Regions Series
Edited by Doris Wastl-Walter
University of Bern, Switzerland

In recent years, borders have taken on an immense significance. Throughout the world they have shifted, been constructed and dismantled, and become physical barriers between socio-political ideologies. They may separate societies with very different cultures, histories, national identities or economic power, or divide people of the same ethnic or cultural identity.

As manifestations of some of the world's key political, economic, societal and cultural issues, borders and border regions have received much academic attention over the past decade. This valuable series publishes high quality research monographs and edited comparative volumes that deal with all aspects of border regions, both empirically and theoretically. It will appeal to scholars interested in border regions and geopolitical issues across the whole range of social sciences.

The Politics of Good Neighbourhood
State, Civil Society and the Enhancement of Cultural Capital
in East Central Europe
Béla Filep

European Borderlands
Living with Barriers and Bridges
Edited by Elisabeth Boesen and Gregor Schnuer

Ethnicity, Gender and the Border Economy
Living in the Turkey–Georgia Borderlands
Latife Akyüz

Community, Change and Border Towns
H. Pınar Şenoğuz

For a full list of titles in this series, please visit
www.routledge.com/geography/series/ASHSER-1224

Community, Change and
Border Towns

H. Pınar Şenoğuz

LONDON AND NEW YORK

First published 2019
by Routledge
2 Park Square, Milton Park, Abingdon, Oxon OX14 4RN

and by Routledge
52 Vanderbilt Avenue, New York, NY 10017, USA

First issued in paperback 2020

Routledge is an imprint of the Taylor & Francis Group, an informa business

British Library Cataloguing-in-Publication Data
A catalogue record for this book is available from the British Library

Library of Congress Cataloging-in-Publication Data
A catalog record has been requested for this book

ISBN 13: 978-0-367-66632-3 (pbk)
ISBN 13: 978-0-8153-5884-8 (hbk)

Typeset in Times New Roman
by Sunrise Setting Ltd, Brixham, UK

To Nejat
(Paramaz Kızılbaş)
and all the beautiful people
who fought without borders for humanity
whose imagination had no bounds

Contents

Illustrations

Figures

Maps

Acknowledgments

The book draws on my doctoral study presented to the Department of Sociology of Middle East Technical University in Ankara. I would not have been able to finish it without the support of many people from whom I found strength and determination. I am greatly indebted to my advisor Ayşe Saktanber for her sustained understanding and support, and her guidance with the fieldwork. From her, I have also learned to do justice to one's own intellectual endeavors and not to let oneself yield to the discouragements. I am also thankful to her for opening up new questions and arousing my interests for further research with her valuable comments about my study. In her coordination, the study received support from the department of Scientific Research Projects in Middle East Technical University Grant No: BAP-07-03-2011-109. I would also like to thank Umut Beşpınar, H. Deniz Yükseker, Helga Rittersberger-Tılıç and Suavi Aydın, who sat in the defense committee, for their comments which helped to crystallize my arguments.

I owe gratitude to many people who facilitated my field research. Hülya Saygılı not only provided me with the most comfortable conditions for a longer stay, but she became a friend and companion with whom I could share many memories about life in Kilis town. Sıdıka Bebekoğlu was a sister and companion who generously shared her knowledge and gave support. Town inhabitants, whose names I cannot list here, showed hospitality, offered solutions when I tackled the problems of daily life during my stay and gave support to my research by sharing their networks. I am thankful to all.

I am grateful to my friends Başak Can, Pınar Yanardağ, Esra Baloğlu, İlker Kabran and Figen Işık. Friends from the Sociology Department kept my academic commitments alive. Among them, I owe particular gratitude to Selda Tuncer. Without her, I would not be able to come through the rigors of scholarship. I am also thankful to my parents, brother and sister-in-law for their endless support. I would like to conclude with special thanks to my life partner Serhat and our six-year-old Vera. They are my 'home' since the failing peace prospects and warmongering on the rise in my country forced us to move away without a chance yet to revisit. I found love and compassion at home whenever I grappled with emotional setbacks. Vera's persistent cheer and sense of pride for this book, though her dream to include her paintings in the book did not come true, was my greatest source of motivation.

Introduction

Borders demarcate nations. They produce belonging. Borders delimit our worlds and give meanings. They define our places in society, to where and whom we feel attached. But they also offer challenges. We push ourselves beyond our borders and seek to free ourselves from limitations, to change our lives in a better direction or at least manage a life that we think is worth living. This book aims to highlight the place of borders in our lives, by pointing to an actual border setting at the southeastern margins of Turkey. I present the story of the transformation of Ottoman Kilis into a border town, as told from the perspective of its dwellers. Drawing on historical and anthropological approaches, I explore the processes and dynamics that transformed the place into a border town and underline its historicity.

I focus on the impact of border on the cultural and economic landscape of Kilis, a town which has been identified as nowhere in Bilad al-Sham,[1]1 but as a border town after the establishment of the Republic, inconspicuous for a long time, yet it recently gained prominence with the eruption of the Syrian crisis in 2011 as opposed to other Eastern and Southeastern border towns. The overriding question of the book concerns the ways in which the border influences the life prospects of town dwellers. Border affects socio-economic strata by introducing new territorial, cultural, and economic barriers that they have to accommodate. The first set of questions concerns the ways in which the border shapes the living of socio-economic strata by severing and shifting their cross-border ties. In what ways do border-induced changes such as loss of land properties, shifting routes of trade and status of goods crossing the border, binational relations of kinship affect families at the border? How do these changes alter the class and status relationships among the strata themselves?

Yet, this book is structured around an overriding question that allows room for the structure (border) and agency (dwellers) dialectics. The impact of the border on dwellers from various strata is not only constraining, but it is also enabling them. This book adopts a framework that inquires about the impact of border on dwellers as well as dwellers' capacity to manipulate and circumvent it, providing room for portraying them beyond being mere victims of state policies. It argues that dwellers can utilize the border as an economic resource, by seeking shelter across and evading legal liabilities to the state and its territorial control. Thus, the main objective of the thesis is to recognize and make visible the struggles and adaptations of dwellers to the border.

While border introduces new distinctions, it also offers new opportunities of encounter and contact across the border. Therefore, I argue that the second set of questions that should be asked must highlight the ways in which socio-economic strata adjust to the border by developing new networks and connections. What sorts of mechanisms and linkages do they develop to rely on? How do they manipulate and circumvent the territorial as well as cultural and economic barriers? In what ways do these shifts in the cross-border ties contribute to the social stratification structure by redrawing class boundaries?

A town is not just designated a border town by being adjacent to the border but as it is argued by Buursink, "it also came into existence because of the border" (Buursink, 2001, pp. 7–8). The processes and dynamics at the border regulate the border-crossing of people, animals, and goods in ways to enable their movement or halt it and, thus, produce different mobilities and enclosures for socio-economic strata, which they have to accommodate in various ways. I assume that movement/ enclosure at the border inflicts its dwellers with values, practices, and relationships that are not found anywhere else in the nation-state and creates "the border experience" (Martinez, 1994) that is often marginalized by social analyses.

Through the lens of ethnographic research and oral history, I explore belonging and social mobility among various socio-economic strata within the context of transition from Ottoman rule to the Republican regime because it will help me to reveal the impact of border on the cultural and economic landscape of Kilis. The aim of the book is to account the social change with the shift of Kilis, an Ottoman inland frontier and a resettlement area for Arabic, as well as Kurdish and Turcoman tribes accused by the Ottoman central power of "political banditry" (Soyudoğan, 2005) into a modern nation-state border. For substantiating my research question, I introduce a theoretical framework that explores the social change in a border town at the intersection of local, national, and transnational processes. I suggest that the border perspective will move the analysis beyond "methodological territorialism" (van Schendel, 2005) that encapsulates the social analysis into the idea of nation-state as unit of analysis. From the vantage point of border, I will demonstrate that social reproduction and social mobility strategies of families in Kilis are incorporated in the broader cultural, social, and economic transformation of Turkey long before globalization started to put limitations to the state sovereignty and territorial control.

Porous border, strangled dwellers

Kilis border town, a small province in southeastern Turkey, has become known worldwide lately due to the extensive news coverage of the Syrian conflict by the world media. Located at the Turkish-Syrian border and presently contiguous to a territory of war and mass demolition, the name of this border town, as well as other towns on the border, probably circulates among United Nations (UN) staff, international non-governmental organization (NGO) employees, government officials, and state bureaucrats. When a flow of migrants along the Turkish border started, the prefabricated camp in Kilis located at the zero point on the border was turned into a showcase for the Turkish government in order to present to the world that it does whatever it can to take care of the human crisis and to accommodate its Syrian guests.[2] The names who paid initial visits to Kilis camp for Syrian migrants in 2012 include UN special envoy for Syria and former UN general secretary Kofi Annan, UN goodwill ambassador Angelina Jolie and the then Prime Minister Recep Tayyip Erdoğan. The visit of Erdoğan seemed to target the attention of world politicians concerned with the plight of Syria as he gave a speech to the migrants, heralding them their victory against the Assad regime in Syria and warning Bashar al-Assad (Kibritoğlu & Çelebioğlu, 2012).

In contrast to the global attention transforming Kilis into a hub of movement, the incoming Syrian migrants simply accentuate for the town dwellers their sense of containment at the border. A local man picturesquely explains it to me by imagining the town as a place squeezed between two stakes at the opposite corners of the town's administrative boundaries with its adjacent city Gaziantep in the north-northwest. He refers to two villages, both named Kazıklı, meaning literally "with pales" in order to portray the town as a territory nipped from its two corners as if impaled and made into a confined place because once you enter the town's boundaries, you have nowhere to go because of the enclosing border with Syria. He says: "We are squeezed between two Kazıklıs."

The increase of "global flows" shapes every corner of the world but in different ways. Globalization creates new spatial inequalities that highlight class disparities beneath them despite its promise towards a borderless world. Although the world is shrinking through a web of communications and exchanges, and geographical movements across the globe seem to defeat the territorial borders of nation-states, there are a great number of people with feelings of confinement and immobility at border geographies in an increasingly globalizing world. Border scholars Hillary Cunningham and Josiah Heyman suggest that the theme of mobility is intrinsic to the questions of power, justice, and inequality (Cunningham and Heyman, 2004, p. 294). For them, mobility, as well as enclosure are "vital to the exploring relationships of differentiation across space" (Cunningham and Heyman, 2004, p. 295). I believe that these points are illustrated in the above-mentioned feeling of being squeezed, expressed by the local man in Kilis.

I have carried out field research in Kilis town between January 2011 and June 2012. The Syrian uprising that broke out in March 2011, three months after I started my field research, had been a challenge for me that made me think more about the sense of containment prevalent among Kilis dwellers. The dwellers depicted the town as devoid of industrial investment and vibrant cultural life, especially after the urban elite had deserted. The border is portrayed as a physical hindrance, an obstacle to the industrial development of the town because, without infrastructure and good highway connections to its surrounding ports and business clusters, it lacks the viable conditions for attracting investment. Kilis was for the dwellers a socio-economic and cultural margin. Culturally, town dwellers have as much attachment across the border as they have with their co-nationals.

Economically, cross-border trade can be a viable option for livelihood and a better life and yet it puts them in jeopardy of criminalization.

As the Syrian opposition grew larger in their struggle to depose of the Assad regime, I bewilderedly kept a close watch on the reactions of local residents. While the national media coverage highlighted the armed conflict and civilian deaths, the dwellers of Kilis continued to argue that the news included exaggeration and there were no such grave conflicts. Eventually, the arguments get varied in that the protests went beyond purpose and the son Assad was actually a good leader, that imperialist forces wanted to create trouble in order to realize their Great Middle East Plan and that Turkey is being trapped by these forces. The prime minister's statement that the Syrian unrest was their internal affair aggravated local fears that the government's attitude would hinder cross-border trade, which was the source of livelihood for a significant number of dwellers. Hence, the incoming of Syrian migrants as well as solidarity actions by local and international NGOs with Syrian opposition in Kilis was met with the resistance of the town community.

Soon after I finished my field research and left the town, the Turkish government closed its border with Syria for security reasons and banned the exit of Turkish citizens. In the Kilis case, it is possible to observe that the globalization framework created new conflicts and spatial arrangements that the town dwellers now have to accommodate. Thus, the emphasis on the global flows and border crossings highlighting mobility may overshadow how these movements enclose spaces of exchange and livelihoods and produce immobility for certain groups.

Not only trade relations, but also other sorts of cross-border exchanges, relations of trust, straddling forms of living such as cross-border labouring or land tenure, kinship, familial alliance and ethnic affinity historically characterized the Kilis border. Transborder movement was part of everyday life, making the dwellers differ from a broader population that is not located adjacent to an interstate border. Then why are the town dwellers engulfed in a strong sense of containment? Paradoxical though it may seem, the border experiences of dwellers cannot be understood without superimposing their sense of containment with the porosity of the Kilis border. This is a question worth thinking about, indicating how volatile and vulnerable life at the border could be.

Historical and ethnographic context

Prior to the Syrian war, Kilis was a border town with a population of about 130,000 located on the Turkish-Syrian border, hosting ethnically Turcoman, Arab, and Kurdish groups with kinship bonds extending to the Syrian side. In the Ottoman Empire, Kilis was an inland frontier for these nomadic tribes, sedentarized by the central authority since the late seventeenth century (Kasaba, 2004). Kilis region also marked the beginning of the Arab provinces of the Ottoman Empire, as distinguished from the Anatolian lands. While the Turkish-Syrian border was delineated between Turkey and France with the Ankara Agreement of 1921, it has been disputable for the Turkish side since the beginning. According to Güçlü, the political boundaries of Syria did not exist prior to the Sykes-Picot agreement, known as a secret alliance between Britain and France in 1916 to define their spheres of influence (Güçlü, 2006, p. 641). But the boundary line nearly corresponded to the assumed linguistic and national boundary of the Turkish area, though it was far northern at its initial demarcation.

When the deputy Ali Cenani from Ayıntap (Gaziantep) made his speech in 1922 on the disputed borderline, he addressed the squeezing of Kilis town by the demarcation of the border. "The line passing just by five kilometer south of Kilis town has rendered that country almost paralyzed," he said (TBMM, 1922, p. 425). The border demarcated in 1921 had torn apart the former Ottoman land from the emerging Republic of Turkey and abandoned north Syria, as well as some of the Kilis agrarian fields to the French Mandate. The provisional delineation of the border was substituted in the meantime by a frontier better suited to the needs of the Turkish side and special regulation of border crossing allowed large landowners to have access to their propriety at the Syrian side. The final delimitation of the border could only be completed with the restoration of Antioch to Turkey in 1939. Big landowning families could continue to claim proprietorship over their lands left at the Syrian side and harvest their crop until their lands were confiscated by the Syrian Baath Party rising in power in 1954 within the context of agrarian reform.

However, the Turkish-Syrian border remains highly contentious in terms of international politics. Regarding the border dispute, the issue was never resolved in the nationalist imagination of linguistic unity because of the significant Turcoman population left in North Syria, particularly in Aleppo and its vicinity. On the other hand, the Cold War policies of the 1950s embraced the border and reinforced it with the wiring and mining of the whole boundary line except Samandağ of Antakya. The rise of the Baath Party in Syria as it won a significant number of parliamentary seats accelerated the political tension between the countries. The background to the tension was the US-led Cold War against the Soviet Union and the American quest to build a transnational security organization in the Middle East against Soviet impact (Baş, 2012). The international relationship between Turkey and Syria was caught in the crisis in 1957 and Turkey dispatched troops to its border. Nevertheless, both Turkey and Syria realized that they would not be able to win without allies in a potential close combat and the political tension loosened.

It is not possible to access the documentation about the mining of the border during 1956–1959. The Turkish government declared that mining of the border was intended to stop smuggling, citing a high-profile incident when smugglers shot two customs agents dead (Köknar, 2004). Nevertheless, the smuggling alone does not explain why the mine laying continued for three years, especially taking into account that illegal trade has actually increased in volume and value after the mining of the border. Moreover, more mines were laid again after the 1980 coup (Özgen, 2010). Mine laying is responsible for many deaths and maimings at the Kilis border.

The Kurdish question and water dispute are two other major sources of political tension. The water dispute goes back to the 1970s as Turkey started the construction of water dams within the context of the Southeastern Anatolia Project (GAP) (Olson, 1997). After Abdullah Öcalan, the leader of the Marxist-Leninist insurgency PKK (*Partiya Karkerên Kurdistan*, Kurdish Workers' Party), found shelter in Syria in the late 1970s, the two issues were played by the countries as trump cards to each other. Yet, Turkey's concern with securing its border against the 'dangerous population' actually dates back to the 1920s and 1930s, and archival sources show that Turkey is quite disturbed with the settlement of deported Armenians, insurgent Kurdish as well as Assyrian and Yazidi emigrants along the Syrian border with Turkey (Altuğ and White, 2009). The Kurdish question between Turkey and Syria found a resolution with the signing of the Adana agreement in 1998 and led Syria to deport the Kurdish leader (Aras and Polat, 2008). But, the emergence of the Kurdish movement in northern Syria and the 2013 declaration of autonomous zones by the Democratic Union Party (PYD) seems to resurrect Turkey's security concerns at its border.

In the light of this brief historical overview, I suggest that international relationships and state politics make the Turkish-Syrian border highly unsettled. I do not only refer to the impact of international politics, but also to the spillovers of internal politics of the states respectively on these borderlands. Despite the ruling Justice and Development Party's (AKP) efforts to mend the fences with Syria until a popular uprising broke out in this country, the relationships between Turkey and Syria have been characterized by boundary disputes, political tensions, and alienation, as well as political negotiations, cooperation seeking, and diplomatic bonds. As part of the Middle Eastern geography, the Turkish border with Syria is characterized by authentic multiculturalism (Doğruel, 2013), ethnic hostility, and religious sectarianism created by nation-state context (Altuğ, 2002), political geographies of ethnic and political divisions (Tuncer-Gürkaş, 2014) and a cosmopolitanism based on multiple and controversial ways of remembering and forgetting the Armenian genocide of 1915 (Biner, 2010).

Local discourses and narratives in Kilis town draw on a historical ethno-symbolism (Canefe, 2002) that utilizes myths of origins and ancestry of Turkish people and memories of a distinct Muslim Anatolian society. The Armenians of Kilis were deported following the 1915 Deportation Law and the Jewish community had completely left by the late 1960s, possibly following the Arab-Israeli conflict in 1967. Local historical accounts emphasize the Turkification and conversion to Islam in the region.[3] These accounts point to the suppression and assimilation of ethnic and religious identities to the national unification and erasure from the popular imagination of the cultural and historical legacy leftover from the Christian communities.

The ethno-religious assimilation is also apparent in the cultural domination of Sunnite Islam, alienating the Alawite identity. The assimilation of Alawite Turcomans in the Kilis region has its historical roots in their settlement in the mid-nineteenth-century Ottoman Empire. İlbeyli Turcomans of Kilis are among the tribes converted to Sunnite Islam with strong tribal bonds and identification due to their relatively recent settlement (Aydın, 2011, p. 11). Turcoman villages are located eastwards, where Kilis lowlands are converging with the Turcoman-dominated plain of Gaziantep, though these lowlands are studded with Kurdish and Arab populations as well. On the other hand, *Kurd Dagh*, a highland border region in eastward Kilis, is predominantly populated with the Kurdish population.

Hence, the case of Kilis affirms the argument that the border is not a homogeneous unit and there can be many borderlands along a geopolitical frontier with particular and distinct border cultures (Donnan and Wilson, 1999). This borderland is distinguished from other places along the Turkish-Syrian border as a strong case of assimilation and stability in terms of the ethnic and political conflicts surrounding it. These features of Kilis borderland make the struggles and adaptations of local dwellers to sustain their cross-border ties in the Middle Eastern political geography even more significant. The refugee arrivals do not undermine these features but rather enhance them. The Syrian refugees, as new inhabitants of the town, have soon become integral to these struggles and adaptations.

Fieldwork on the border

Borders, as margins of the nation-state, are marginalized in official and popular discourses. Depicted as entrenched in the traditional community relationships and underdevelopment, the borders are stigmatized as sites escaping the state order and open to corruption and illegal practices. The borders are also stigmatized by border dwellers. In Kilis, for instance, most of the dwellers associate the border with its unregulated trade and illicit economy. Kilis dwellers notoriously name their efforts to earn a livelihood at the border as "smuggling" and place it at the center of their border perceptions, as if it is the only distinctive characteristics of this border. To put it differently, borders bring their own ideologies, stories, and ways of interaction that the border researcher must delve into in order to explore what living at the border meant to the border dwellers. This might prove rather challenging when this is not legible in the official and archival records, sources of information are not cooperative, or secrecy dominates among the dwellers.

The research consists of 18-months' fieldwork on belonging and social mobility processes among various socio-economic strata in Kilis. Anthropologists classify ethnography as a genre of storytelling, reminding us that our ethnographic works are also narratives transforming the sense of fragmentation, contingency, and dislocation experienced in the field (Bruner, 1986; Rapport, 2000). This book locates the agency of town dwellers in accommodating themselves to the border on the foreground against the view of border dwellers as being mere victims of artificial territorial boundaries drawn by colonial powers in the Middle Eastern context. I explore different strategies adopted by various socio-economic strata to navigate territorial borders and group boundaries. For this purpose, I rely on ethnographic and oral history approaches as well as a reflexive anthropology establishing an intimate rapport with the town dwellers to overcome the challenges of doing fieldwork on the border.

The border context has imperatives that determine the research question as well as strategy. Kilis, as the provincial hinterland of Ottoman Aleppo before the demarcation of the border (see Map 0.1), remained a moderately big town after it

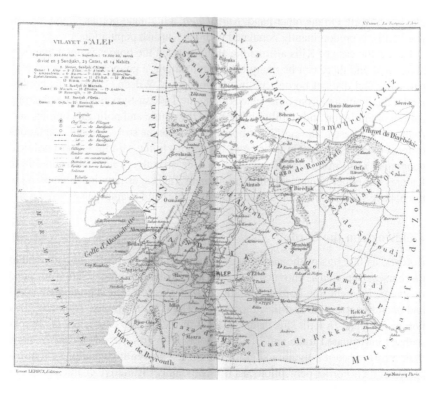

Map 0.1 Cuinet's map showing Kilis as district (*kaza*) of Aleppo province, 1835
          (Retrieved from Wikimedia Commons)

became annexed to the city of Gaziantep. The town bears the traces of a highly
stratified social structure inherited from the Ottoman landholding class and agricultural production. The cultural and architectural inheritance of long-established
notables is still present, though notable families have lost their social and economic power since the 1960s. Even though these families remained among the
wealthy strata of the town, I realized their feelings of decline should be interpreted
as part of their social mobility experiences with the demarcation and consolidation of the border. The historical background that informs identity formation in
this border region indicated that the question of belonging should be addressed
by a genealogical study of family histories through Ottoman archives. Life stories
of notables conduced towards an inquiry about the ways in which social mobility
processes at the border pave the ground for critical belongings and transform the
meanings of illegality, wealth, and work, which are inflicted by a border culture
difficult to penetrate. This is an important reason why I make much use of partic-
ipant observation data.

  As the method of study, I rely on oral history and ethnographic research supported
by in-depth interviews, archival research, as well as trope analysis. Embracing oral

history within the ethnographic context is not uncommon. Yet, it extends the question of intersubjectivity, that anthropological writing tackles since the 'interpretive turn' (Vila, 2003) after *Writing Culture* (Marcus and Clifford, 1986), to oral history interviews and renders oral history and ethnographic study alike (Di Leonardo, 1987). However, the added value of oral history interviews, in contrast to in-depth interviews, is that oral history allows the researcher to unravel the past through a multiplicity of standpoints and encompasses the complexity of conflicts (Thompson, 1998). Oral history reveals the silenced and omitted accounts of the past, often shrouded in secrecy by the prevailing modes of history writing, an impact that could be observed not only in the narrating of family histories among Kilis notables but also in the peasants' accounts of their relationships with the notables.

In the total, I conducted 44 recorded (voice recording and note-taking) interviews, including more than one person in a few cases. The interviewees were recruited from a social composition including three socio-economic strata: namely, old wealth, new wealth and middle class, and rural and urban poor (see the Appendix for Interviewee Profiles and Explanatory Notes for Oral History and In-depth Interviews). I define old wealth as traditional landed and trade notables, a status distinction which I discuss in length in Chapter 2 on the notables. New wealth and the middle class are comprised of extended families that are usually—but not necessarily—from a rural background and have moved up among the ranks of the urban middle class. One shortcoming is that I could not recruit interviewees among the small circle of new wealth, who have started to make investments in Istanbul since the mid-1960s and settled there all together after the 1980 coup. I try to meet this deficit by giving a profile of new wealth in Istanbul, by introducing the story of a barber in Chapter 4 on new wealth. Members of new wealth in Istanbul and extended families from a rural background living in Kilis experienced upward mobility through the same mechanisms and wealth-generating practices during the 1960–1980 period of import-substituting industrialization. The new middle class also includes heirs of old notables, whose social reproduction strategies are based on education and urban employment. Finally, I classify the rural and urban poor as lower strata. These interviewees are usually the offspring of independent peasant families or families working at the landlord's household as domestic and casual labourers.

The study recruits a wide arrange of interviewees among local notables, chieftains, tradesmen and shopkeepers, professionals and intellectuals, border villagers, and urban poor. They also include a few former mayors and parliamentarians, which contributed to my understanding of how local political dynamics articulated with the broader border politics. Interviews with officers at governmental and semi-public institutions added to the information about the socio-economic structure and dynamics of social change. Regarding the sectoral composition, the background of interviewees mainly covers agriculture, trade, and transport sectors. When I could not conduct a recorded interview, the interactions and encounters in the field kept a vigilant eye to subaltern voices. These range from casual labourers and unemployed youth to women peddlers and border villagers. I had informal dialogues with groups of high school students, women gatherings, and border villagers.

There were many cases where I could not take personal information such as age and current profession. But the dialogues gave me important clues about the interviewees' position and status as well as insights about the history and facts about illegal border trade and its impact on the socio-economic structure of the city. Thus, I prefer to give the birth year intervals of five years in order to classify the interviewees along the age groups. Regarding the age scale, old wealth typically included the oldest age groups, with birth years ranging between 1920 and 1955, with the exception of Latife born in 1911. The birth year intervals ranged between 1930 and 1975 for new wealth and the middle strata as they have been interviewed principally about the shadow economy of the import-substituting industrialization period of 1960–1980, except an old-generation woman who could give information about her father's involvement in smuggling livestock as early as the 1940s and a younger transport company owner. The age distribution among the rural and urban poor is almost even as the interviews were concentrated both on the agrarian labour structure between the 1930s and 1960s, and the cross-border trade introduced in the mid-1990s. The book draws much of these ethnographic data offered by the encounters and exchanges with these unrepresented groups.

Recorded interviews were likely to take the form of oral histories about social and economic life. As interviewees were reluctant to talk about the present situation, they were more inclined to talk about the past in order to avoid, for instance, revealing their involvement in illegal trade. While urban notables and the new rich gave accounts of their family life and social history of the town, border peasants and the urban poor narrated their struggle with regard to subsistence and the change in their relationship with the landowning class. The border peasants were also afraid of law enforcement and military surveillance at the border.

Since I have met with difficulties in conducting recorded interviews, I have complemented them with snippets of responses (Vila, 2000, p. 254) and micro-narratives (Doevenspeck, 2011) in order to uncover hidden narratives beneath the dominant stereotypes and prejudices and to approach the everyday experiences and practices at the border. Such pieces of conversations are employed by border studies in inquiring the narrative construction of boundaries, identities, and belonging at the border (Paasi, 2003; Vila, 1997, 2000; Doevenspeck, 2011; Pelkmans, 2006; Flynn, 1997). Putting emphasis on narrativity does not necessarily mean to employ a narrative methodology that applies the procedural set of techniques to relate small-scale stories to the wider context of narratives. Rather, I work with local narratives in order to disclose the 'border talk', i.e. in what ways dwellers talk about the border, which issues and problems they bring to the foreground, why these are being told and not the others. As I refer to in the analysis chapters of this book, the border talk in Kilis town included issues about the demarcation of border and local resistance against the French occupation, trade activities connecting Kilis with trade and finance centers in the region, the use of border as economic resources, and cross-border marriages and kin relationships. These narrative accounts reveal the shifts in power relations experienced in everyday life.

Finally, I support my discussion with reference to a purposive sampling on local and national newspapers in order to study the criminalizing discourses and their possible effects shaping the experiences of border dwellers. During the fieldwork, I

Figure 0.1 A prayer break in the border village of *Kurd Dagh* (Photo courtesy of the author)

have regularly tried to follow daily news appearing in local papers. There are four local newspapers in Kilis, among which two of them date back to the early Republican period. Although they tend to sound like tabloid press publishing mainly the news circulated by the local police department of anti-smuggling, they reveal the differences of opinion and established alliances among various positions of political adherence. The websites of these newspapers provide virtual forums for readers expressing comments and, thus, they give clues about the public opinion of the local community. The study also draws on archival sources, including international treaties, laws and regulations about border control and customs as well as smuggling, and minutes of parliament meetings. These documents shed light on the official discourses related to controversial issues of border management, the definition of border as a legal category and the criminalization of transgressive activities.

Short visits to the town after I finished my fieldwork allowed follow-up observations about the impact of the Syrian conflict. These visits included professional occasions in October 2012 and March 2016, when I worked as translator-fixer for MSF and Deutsche Welle respectively. I also paid a week-long visit for research in July 2015 and held semi-structured interviews in the town center and border villages of Elbeyli and Kurd Dagh (see Figure 0.1).

Conceptual framework

Border studies constitute a theoretical background to this study. However, as Paasi (2011) argues, this scholarly literature is not yet an integrated field of study. As the researchers of border, we cannot refer to a 'border theory' as groundedly as we may speak of the state theory or theory of social classes. Border studies has been revived as an interdisciplinary realm of investigation under the impact of global-ization, but mainly adopted anthropological approaches prioritizing conceptual boundaries rather than territorial borders or even embracing a theoretical frame-work that puts the territorial borders out of sight. Moreover, as studies on bor-der are proliferating, the understanding of borders are perpetually changing and diverging depending on the research problems addressed and eventually become part of theorization. Nevertheless, research on border is still empirically compart-mentalized, largely deprived of historical and comparative analyses.

The conceptual framework of my study combines the arguments produced from within border studies and oral history with the debates on post-Ottoman social strat-ification in order to provide a transdisciplinary approach to power, hierarchy, and in/exclusion in a particular border region. Oral history and theories of social stratifi-cation are methodological imperatives of the research as they substantiate historical analysis with the concept of socio-economic mobility among different strata.

Border, frontier, boundary and borderland

The literature review in Chapter 1 is fully devoted to highlighting the analytical power of the concepts adopted in this book with reference to border studies. Still, this section shortly introduces some of these concepts.

In this book, the term "border" is used to refer to the territorial and, in the case of Kilis, political boundary. Border drawn between two nation-states is necessar-ily territorial and cartographic, i.e. delineated on a map. As opposed to the term "frontier", border encloses neighborhood, communities, and nations and turns them into territorially-bounded units. As Strassoldo argues, frontiers are "areas of growth into 'virgin' territories" (Strassoldo, 1980, p. 50). Famously known as Frederick Jackson Turner's frontier thesis, the term frontier is used to denote the territorial expansions of civilizations and empires (Baud and van Schendel, 1997). In other words, frontier does not signify limits or barriers, but openness, expan-siveness and dynamism.

A new genre in border studies employs the notion of "borderland" as a focal metaphor to denote a liminal space where two distinct cultures face and counter-pose each other (Wilson and Donnan, 1998). The literature review extensively cites the criticisms about the new borderlands genre from within border studies in order elaborate on the approach to "border/land as liminal space." The term "borderland" for this book signifies the "existence and impact of a border on the human landscape" (Newman, 2003, p. 19). It denotes a border location embed-ded within networks straddling across the territorial border between two or more states (Baud and van Schendel, 1997; Wilson and Donnan, 1998). Embracing the notion of borderland reiterates the fact that the border is more than a line bisecting a geographical space. Constantly shifting, Baud and van Schendel fig-uratively describes it as an accordion in order to portray contracting and expand-ing ties radiating at the two sides of the territorial border in response to various contentions and strife raised by state politics and transnationalism (Baud and van Schendel, 1997).

Within the realm of border studies, the term "border" and "boundary" could be used interchangeably. In this book, boundaries are associated with communal borders to "signify the point at which "we" end and "they" begin" (Migdal, 2004, p. 5). Contemporary border studies combine the concepts of borders (territorial delimitations) and boundaries (social categorizations) in a single analytical framework to explore the bordering processes, through which, as Newman argues, the categories of difference or separation are created (Newman, 2003). The source of inspiration for this development is the turn to Barth's paradigmatic ideas on ethnic boundaries (Donnan and Wilson, 1999). Instead of defining ethnic identity as a set of features shared by the member of groups, Barth (1969) looked at the inclusion and exclusion processes through which ethnic boundaries are maintained and re-confirmed. Borrowed from Barth, boundaries acquire an analytical power to study apparently unrelated phenomena of boundary-making within the nation and the territorialization and policing of borders enclosing the nation, recently illustrated by the analyses of Fassin (2011), Parizot (2008), and Yuval-Davis et al. (2017), among many others.

In this book, boundaries point to the group (class) identifications among socioeconomic strata in Kilis. To set it more clearly, this book valorizes Barth's original question about how the group identifications could be maintained despite their boundaries being transgressed. It shows how the tensions of old/new wealth and upper strata/peasantry are constantly negotiated through marriages and other forms of alliances across borders and boundaries.

Social mobility

I draw on an oral history approach in order to underline my methodological premises related to social mobility. Social mobility can be defined as the movement up and down the stratification (Kerbo, 2006). Border historians turn to oral history in order to "reconstruct the historical self-images and perceptions of social groups in the borderland and the impact of these on people's political, economic, and cultural behavior" (Baud and van Schendel, 1997, p. 242). I adopt oral history as a theoretical perspective to inform social stratification analyses with particular emphasis on social mobility, which is brilliantly illustrated by the works of oral historians Daniel Bertaux and Paul Thompson (1997). They argue that narrative approach exceeds the often-preferred survey method in social mobility studies by revealing the crucial importance of local context. While social mobility processes in Kilis town could be analyzed in conformity with regular patterns of mobility, they critically diverge in terms of subjective perceptions and evaluations and local structures of opportunities. Narrativity adds great power in explaining the reproduction, transformation, emergence, or disappearance of class and status in border zones, where a fast buck could be as easily spent as it is earned.

The overly changing nature of borders and border policies differentially allowing and hindering the border-crossings generate a context of unequal power relations in which the social hierarchies of class, status, and prestige fluctuate. Arbitrary border policies allocate unequal risks to various strata but they normalize them as consequences of living at the border (Cunningham and Heyman, 2004, p. 294). Therefore, this book adopts an analysis of social stratification structure with the aim of determining the parameters of social mobility rather than describing the class positions and identities. An oral history approach to social mobility helps to unravel family histories, which are "such extraordinarily rich sources of hard information directly relating to the construction of social trajectories [of social mobility]" (Bertaux, 1989, p. 85). According to Bertaux, the key concept to the analysis of social mobility is *transmission*.

An oral history approach relies on case histories of families in order to trace the social status as construed as a property of family groups, which is transmitted among generations (Bertaux and Bertaux-Wiame, 1997). Rather than individuals, parent–offspring relationships are inquired. Bertaux-Wiame and Thompson (1997) define social mobility in terms of regular patterns such as intergenerational occupational transmission, the role of marriage, access to education and housing as cultural capital. In the case of Kilis, the familial histories particularly account for the local notables that depends on their patrimonial wealth as well as paternalistic relationships of agrarian production. However, they also shed light on what ways new wealth as well as the lower strata diverge from expected social trajectories of class reproduction, i.e. in cases that the occupation is not transmitted between generations. While these patterns constitute the vertical mobility paths of families by transmitting their economic and cultural capital to their offspring, I also add the conversion of capital held in one form to another as an aspect of horizontal mobility.

Conversion of one type of capital into another, Bourdieu argues, occurs only if the latter is "more profitable or more legitimate in the current state of instruments of reproduction" (Bourdieu, 1998, p. 277). For Bourdieu, there are four types of capital, including economic, cultural, social, and symbolic. Yet he considers the economic and cultural capital possessed by individuals as the most important. Bourdieu emphasizes that cultural capital is heritable and critical to the reproduction of social standing among individuals. This type of mobility, usually neglected by other models of social mobility, can also shed light on various conflicts between various groups among social strata (Weininger, 2005).

Social stratification

A transdisciplinary study of in/exclusion in Kilis town needs to be framed within the context of local power and inequality structures that remain latent in border zones. Although the cartographic mapping makes the territorial boundaries appear as crossing deserted plateaus and mountains, wide and barren fields without human trace, borders are not demarcated on blank space. The demarcation of border shifts the social stratification among inhabitants and alters the forms of identification.

Social stratification refers to a hierarchy and inequality of ranking among social groups in the society. It incorporates institutional processes that create these hierarchies and inequalities as well as mobility mechanisms that shift them. Classification among social groups can be done according to a set of categories: namely, the degree of inequality, rigidity of the stratification system, ascriptive traits and degree of status crystallization (Grusky, 1994). A Marxian tradition, informed by the theories of Marx as well as Weber, has long produced numerous work to bring the class and status groups within a single analytical framework. Bourdieu's reproduction theory (1984) has made a great leap forward to straddle the division between class and status, by incorporating the notion of distinction and taste and defines classes on the basis of their internalized dispositions—or tastes—rather than objective conditions of existence. While this book adopts Bourdieu's approach to social mobility in terms of reproduction and conversion, it nevertheless embraces the concept of 'strata' as a key concept of analysis in order to address the unresolved division between class and status groupings.

Studying the demarcation of Kilis border town has a theoretical backdrop informed by the Ottoman social stratification system. The latter lies at the roots of class structure of modern Turkey. One approach in the Ottoman historiography conceptualizes the Ottoman social stratification as Asiatic, with peculiar ascription and achievement criteria than those existing in Western Europe. This line of thought conceives the Asiatism as an intermediate formation in the transition of advanced agrarian society (Grusky, 1994, p. 9). For example, Şerif Mardin (1967) is a proponent of this approach, tracing the origins of contemporary class structure in Turkey back to the Ottoman social order. Mardin conceptualizes the Ottoman social order in a dichotomy of the ruling class and the ruled class, the latter mainly being the peasantry. According to him, the most important shift in the Ottoman stratification structure had been the Land Code of 1858 allowing private proprietorship of the state lands. The private proprietorship of land had led to the emergence of a new class of landowners as the ruling class in the provincial Ottoman. An important aspect of Ottoman social stratification was its rigid nature as expressed by the ruling class' concern for "everyone keeping his proper place" (Mardin, 1967, p. 129).

Mardin argues that the nascent Turkish regime has largely preserved the Ottoman social stratification structure split by a historical dichotomy between the ruling and the ruled, as well as intra-group conflict among the ruling elite. The provincial rulers in the Ottoman Empire are regarded by the Ottoman historiography as "notables" playing a role as intermediaries between central authority and the local population (e.g., Hourani, 1968). For these scholars, the feudal relationships, such as large landholding, which are inherited by the Republican State, complicate the analysis of social stratification structure in terms of capitalist classes.

A second line of thought in Ottoman historiography explores the unfolding of Ottoman history on the basis of the mode of incorporation of the Ottoman Empire into the capitalist world economy: peripheralization thesis (Keyder, 1991). This approach, in turn, aims to define social classes in terms of production relationships and ownership of the means of production. The two lines do not only diverge in defining the mode of production (Asiatic or peripheral capitalist) but also contend each other in the definition of social stratification structure in the Ottoman order. Khoury tries to bridge the two lines by discussing the definition of local notables as a class with respect to property or its relationship to the means of production as well as to the social position of its constituents (Khoury, 1990, p. 219). He discusses the notables as a class per se, but also as a "class in formation" by emphasizing the significance of the social standing together with the ownership of the means of production and the patronage relationships, which put to their disposal a wider range of benefits and services than the landownership would. I adopt Khoury's definition in conceptualizing the notables as a social class to unearth their social reproduction and mobility strategies, which do not attract enough interest in the Turkish scholarship.

My discussion about the inner social stratification in Kilis town does not go against the grain of the Turkish social stratification debate but rather engages in a conversation with it. My analysis critically adopts the peripheralization thesis and dwells on the agrarian question that occupied the attention of Ottoman historiography for a long time. Briefly, this thesis argues that the petty-commodity production prevailed among the peasants after the World War II together with the decline of large landholding and sharecropping. I underline the role of local notables, the status of peasantry, and shifting relationships on land tenure as well as patronage in the transition to the Republican regime. I pay particular attention to trace the shift of 'traditional' relationships based on reciprocity and kinship to scrutinize the ways in which these relationships continue to be effective, informing new power and inequality structures in market economy.

I contribute to the debate by shedding light on the mechanisms of illicit wealth generation and redistribution, as well as their normalization embedded in everyday transactions and exchanges, as part of social reproduction and mobility strategies. I demonstrate that the wealth generation through illicit means underpins the vested interests of traditional landed notables and prevents the peasantry from gaining independence. This book elaborates on the historical evolution of underlying mechanisms with the transition from the late Ottoman to the Republican era. These mechanisms have recently come on the radar within the framework of post-1980 trade liberalization and neoliberal governance. Therefore, the rise of new wealth, capitalizing on the post-1980 trade liberalization policies, has mainly attracted scholarly attention as a post-1980 phenomenon (Bali, 2007). However, this book points to the ascendance of new wealth between the 1960s and 1980s. It also shows how new relations of dependency are established between the ascendant wealth and the peasant poor. The analysis of social mobility processes, moving beyond methodological territorialism, captures the sub-national and transnational dynamics of wealth generation.

Presentation of chapters

I elaborate further on the theoretical background of the above-mentioned conceptual sources in Chapter 1 in order to explore the inner stratification and change in Kilis border town. I give an account of seemingly divergent theoretical threads on border studies: first, the conception of borders as liminal spaces and second, the conception of borders as state margins. I evaluate these threads on the basis of their strength and weaknesses in order to redress an interdisciplinary approach to the study of borderlands. Lastly, I review the Turkish literature on border in order to highlight the main themes and questions.

The objective of the three following parts is to present the empirical analysis of social mobility strategies among three strata, each being devoted a separate part. In Chapters 2 and 3, it is argued that subjective perceptions and evaluations of traditional notables about the shift in their status as old wealth have culminated in their experience of falling from grace since the 1960s. By carrying out a discussion about whether the local notables could be considered as early capitalists, these chapters indicate that these families could benefit from new opportunities in the circulation of goods and gold as the border transit regime allowed them to access their landed estates and the trade markets in northern Syria and yield an economic accumulation. Thus, they show that local notables had to embrace illegal means of economic accumulation and accept what was once disgraceful for them in order to reproduce their social standing, while this process irrevocably undermined the social capital that traditional landed notables relied on to distinguish themselves from trade notables, the latter being more adept in seizing new opportunities of illicit trade.

Chapters 4 and 5 focus on the growth of the shadow economy along Kilis border and explore the social and economic conditions of the rise of new wealth since the 1960s. They detail how the town dwellers took benefit of the protective measures during the import-substituting industrialization period of 1960–1980 by reckoning rents to the illegal entry of consumer goods as well as gold and foreign currency. Chapter 5 shows that the illegal trade of gold as well as consumer goods promoted upward mobility by undermining structural constraints of social stratification and shifted the social and urban landscape of the town by embedding the local economy into national and transnational shadow networks.

The third section is the last one devoted to the empirical analysis of the ways in which rural and urban poor turned into border labourers and normalized their engagements in illegal trade. Chapter 6 highlights the historical conditions in which the rural poor emancipated themselves from the paternal relationships of patronage established with large landholders, which, yet, were quickly replaced by large-scale entrepreneurs that drew on traditional power structures based on land tenure and kinship. It argues that illegal trade at the border gave rise to a border economy, with the mining of the border from the mid-1950s. Then it explores how the cross-border trade regulations officially introduced in Kilis border in the mid-1990s permitted the town dwellers to yield differential profits and reckon upon a semi-legal small-scale trade. Chapter 7 demonstrates that the rural and urban poor, in the absence of patronage by large-scale entrepreneurs, are increasingly exposed to the dangers of criminalization and being unable to pay criminal fines, dangers of growing economic indebtedness. It also points to the ways in which the poor historically relied on kinship relations to normalize their exchange practices deemed illegal by local authorities.

The conclusion highlights the main findings by drawing on the overriding debates of the book with particular emphasis on social trajectories of socio-economic strata and their stories of social mobility and belonging. Finally, the

Epilogue to this book offers a snapshot of the post-conflict situation in this borderland with a focus on the demographic change, local hospitality towards refugees, emergence of a humanitarian–military nexus and shifts in the local labour market after the refugee arrivals, as well as the ascendance of a war economy and new trade geographies in the broader region.

Notes

1 Bilad al-Sham historically refers to the geographical region, which is also known as the Levant or Greater Syria.
2 The governmental policy stipulates the temporary protection of Syrian refugees and does not provide the right to asylum. This topic will be revisited in the Epilogue.
3 For instance, according to Ekrem, a local historian, the Turkification of Kilis region was due to a Karahanlı prince as he defeated the Byzantine dominion and the settlement of his retinue of Turcoman tribes, while the conversion of the locals to Islam happened under the Ottoman rule after the conquest of these lands with the Mercidabık War in 1516. But, Aydın (2011) argues that the Turcoman tribal identity was originally quite dissimilar to the modern Turkish ethnicity.

References

Altuğ, S. (2002) *Between Colonial and National Dominations: Antioch under the French Mandate (1920–1939)*, unpublished MA thesis, Boğaziçi University, Istanbul.
Altuğ S. and White B. T. (2009) "Frontières et pouvoir d'État. La frontière turco-syrienne dans les années 1920 et 1930", *Vingtième siècle*, 103: 91–104.
Aras, B. and Polat, R. K. (2008) "From Conflict to Cooperation: Desecuritization of Turkey's Relations with Syria and Iran", *Security Dialogue*, 39(5): 475–495.
Aydın, S. (2011) "Baraklar: Antep'in İskan Halkı", pp. 155–234 in M. Nuri Gültekin (ed.), *Ta Ezelden Taşkındır: Antep*, İstanbul: İletişim.
Bali, N. R. (2007) *Tarz-ı Hayat'tan Life Style'a: Yeni Seçkinler, Yeni Mekanlar, Yeni Yaşamlar*, Istanbul: İletişim Yayınları.
Barth, F. (1969) *Ethnic Groups and Boundaries: The Social Organization of Culture Difference*, Boston: Little, Brown.
Baş, A. (2012) "1957 Suriye Krizi ve Türkiye", *Historical Studies*, 4(1): 89–109.
Baud, M. and van Schendel, W. (1997) "Toward a Comparative History of Borderlands", *Journal of World History*, 8(2): 211–242.
Bertaux D. and Bertaux-Wiame, I. (1997) "Heritage and Its Lineage: A Case History of Transmission and Social Mobility over Five Generations", pp. 62–97 in I. Bertaux-Wiame and P. Thompson (ed.), *Pathways to Social Class: A Qualitative Approach to Social Mobility*, Oxford: Clarendon Press.
Bertaux, D. (1989) "From Methodological Monopoly to Pluralism in the Sociology of Social Mobility", *The Sociological Review*, 37(1): 73–92.
Bertaux-Wiame, I. and Paul T. (1997) "The Familial Meaning of Housing in Social Rootedness and Mobility: Britain and France", pp. 124–182 in I. Bertaux-Wiame and P. Thompson (eds.), *Pathways to Social Class: A Qualitative Approach to Social Mobility*, Oxford: Clarendon Press.
Biner, Ö. Z. (2010) "Acts of Defacement, Memory of Loss: Ghostly Effects of the 'Armenian Crisis' in Mardin, Southeastern Turkey", *History & Memory*, 22(2): 68–94.
Bourdieu, P. (1984) *Distinction: A Social Critique of the Judgment of Taste*, Cambridge: Harvard University Press.

Bourdieu, P. (1998) "Forms of Power and Their Reproduction", pp. 263–299 in *The State Nobility: Elite Schools in the Field of Power*, Oxford: Polity Press.

Bruner, E. M. (1986) "Ethnography as Narrative", pp. 139–158 in Victor M. Turner and E. M. Bruner (eds.), *The Anthropology of Experience*, University of Illinois Press.

Buursink, J. (2001) "The Binational Reality of Border-Crossing Cities", *GeoJournal*, 54: 7–19.

Canefe, N. (2002) "Turkish Nationalism and Ethno-Symbolic Analysis: The Rules of Exception", *Nations and Nationalism*, 8(2): 133–155.

Clifford, J. and Marcus, G. E. (1986) *Writing Culture: The Poetics and Politics of Ethnography*, University of California Press.

Cunningham, H. and Heyman, J. (2004) "Introduction: Mobilities and Enclosures at Borders", *Identities: Global Studies in Culture and Power*, 11: 289–302.

Di Leonardo, M. (1987) "Oral History as Ethnographic Encounter", *The Oral History Review*, 15(1): 1–20.

Doevenspeck, M. (2011) "Constructing the Border from Below: Narratives from the Congolese-Rwandan State Boundary", *Political Geography*, 30: 129–142.

Doğruel, F. (2013) "An Authentic Experience of 'Multiculturalism' At the Border City of Antakya", *Çağdaş Türkiye Tarihi Araştırmaları Dergisi/Journal of Modern Turkish History Studies* XIII/26(Bahar/Spring): 273–295.

Donnan, H. and Wilson, T. M. (1999) *Borders: Frontiers of Identity, Nation and State*, Oxford: Berg.

Fassin, D. (2011) "Policing Borders, Producing Boundaries: The Governmentality of Immigration in Dark Times", *Annual Review of Anthropology*, 40: 213–226.

Flynn, D. K. (1997) "'We Are the Border': Identity, Exchange, and the State Along the Benin-Nigeria Border", *American Ethnologist*, 24(2): 311–330.

Grusky, D. B. (1994) "The Contours of Social Stratification", pp. 3–35 in David B. Grusky (ed.), *Social Stratification: Class, Race, and Gender in Sociological Perspective*, Boulder: Westview Press.

Güçlü, Y. (2006) "The Controversy over the Delimitation of the Turco-Syrian Frontier in the Period between the Two World Wars", *Middle Eastern Studies*, 42(4): 641–657.

Hourani, A. (1968) "Ottoman Reform and the Politics of Notables", pp. 41–68 in W. R. Polk and R. L. Chambers (eds.), *Beginnings of Modernization in the Middle East*, University of Chicago Press.

Kasaba, R. (2004) "Do States Always Favor Stasis? Changing Status of Tribes in the Ottoman Empire", pp. 27–48 in J. Migdal (ed.), *Boundaries and Belonging*, Cambridge: Cambridge University Press.

Kerbo, H. R. (2006) "Social Stratification", pp. 228–236 in C. D. Bryant and D. L. Peck (eds.), *21st Century Sociology: A Reference Handbook*, Thousand Oaks, CA and London: Sage Publications.

Keyder, Ç. (1991) "Introduction: Large-Scale Commercial Agriculture in the Ottoman Empire?", pp. 1–16 in Ç. Keyder and F. Tabak (eds.), *Landholding and Commercial Agriculture in the Middle East*, Albany, NY: State University of New York.

Khoury, P. S. (1990) "The Urban Notables Paradigm Revisited", *Revue du monde musulman et de la Méditerranée*, 55–56: 215–230.

Kibritoğlu, M. & Çelebioğlu, R. (2012) "Başbakan Erdoğan'dan Esad'a sınırdan gözdağı", Hürriyet, 6 May. Available at http://www.hurriyet.com.tr/gundem/basbakan-erdogandan-esada-sinirdan-gozdagi-20496557 [accessed at 7.5.2018]

Köknar, A. (2004) "Turkey Moves Forward to Demine Upper Mesopotamia", *Journal of ERW and Mine Action*, 8(2): 54–55.

Mardin, Ş. (1967) "Historical Determinants of Stratification: Social Class and Class Consciousness in the Ottoman Empire", *Siyasal Bilgiler Fakültesi Dergisi*, 22(4): 111–142.

Martinez, O. (1994) *Border People: Life and Society in the U.S.–Mexico Borderlands*, University of Arizona Press.

Migdal, J. (2004) "Mental Maps and Virtual Checkpoints: Struggles to Construct and Maintain State and Social Boundaries", pp. 3–26 in J. Migdal (ed.), *Boundaries and Belonging: States and Societies in the Struggle to Shape Identities and Local Practices*, Cambridge University Press.

Newman, D. (2003) "On Borders and Power: A Theoretical Framework", *Journal of Borderland Studies*, 18(1): 13–25.

Olson, R. (1997) "Turkey–Syria Relations since the Gulf War: Kurds and Water", *Middle East Policy*, 5(2): 168–193.

Özgen, N. (2010) "Mayınlı Arazilerin Temizlenmesi Üzerine", *Almanak*, Istanbul: SAV.

Paasi, A. (2011) "A 'Border Theory': An Unattainable Dream or a Realistic Aim for Border Scholars?", pp. 11–31 in D. Wastl-Walter (ed.), *A Research Companion to Border Studies*, Aldershot: Ashgate.

Paasi, A. (2003) "Boundaries in a Globalizing World", pp. 462–472 in K. Anderson, M. Domosh, S. Pile and N. Thrift (eds.), *Handbook of Cultural Geography*, London: Sage Publications.

Parizot, C. (2008) "Crossing Borders, Retaining Boundaries: Kin-nections of Negev Bedouin in Gaza, West Bank, and Jordan", pp. 58–84 in S. Hanafi (ed.), *Crossing Borders, Shifting Boundaries: Palestinian Dilemmas*, Cairo: American University in Cairo Press.

Pelkmans, M. (2006) *Defending the Border: Identity, Religion, and Modernity in the Republic of Georgia (Culture and Society after Socialism)*, University of Cornell Press.

Rapport, N. J. (2000) "The Narrative as Fieldwork Technique: Processual Ethnography for a World in Motion", pp. 71–95 in V. Amit (ed.), *Constructing the Field: Ethnographic Fieldwork in The Contemporary World*, London, Routledge.

Soyudoğan, M. (2005) *Tribal Banditry in Ottoman Ayntab (1690–1730)*, unpublished MA thesis, Bilkent University.

Strassoldo, R. (1980) "Centre-Periphery and System-Boundary: Culturological Perspectives", pp. 27–61 in J. Gottman (ed.), *Centre and Periphery: Spatial Variations in Politics*, London: Sage Publications.

TBMM (1922) *Gizli Celse Tutanakları* (Minutes of Closed Session of TGNA), Volume 3, Legislative Year III, 15.1.1922.

Thompson, P. (1998) "The Voice of the Past: Oral History", in R. Perks and A. Thomson (eds.), *The Oral History Reader*, London: Routledge.

Tuncer-Gürkaş, E. (2014) "Belirsizlik mıntıkası ya da daimi istisna hali olarak sınır: Güneydoğu kampı içinde Mardin-Kızıltepe ikiz kampları", *Toplum ve Bilim*, 131: 219–235.

van Schendel, W. (2005) "Studying Borderlands", pp. 1–24 in *The Bengal Borderlands: Beyond State and Nation in South Asia*, London: Anthem Press.

Vila, P. (1997) "Narrative Identities: The Employment of the Mexican on the U.S.–Mexican Border", *Sociological Quarterly*, 38(1): 147–194.

Vila, P. (2000) *Crossing Borders, Reinforcing Borders: Social Categories, Metaphors, and Narrative Identities on the U.S.–Mexico Frontier*, Austin: University of Texas Press.

Vila, P. (2003) "Introduction: Border Ethnographies", pp. ix–xxxv in Pablo Vila (ed.), *Ethnography at the Border (Cultural Studies of the Americas)*, University of Minnesota Press.

Weininger, E. B. (2005) "Chapter 4: Foundations of Pierre Bourdieu's Social Class", pp. 119–149 in E. O. Wright (ed.), *Approaches to Class Analysis*, Cambridge University Press.

Wilson, T. and Donnan, H. (1998) *Border Identities: Nation and State at International Frontiers*, Cambridge University Press.

Yuval-Davis, N., Wemyss G. and Cassidy K. (2017) "Introduction to the Special Issue: Racialized Bordering Discourses on European Roma", *Ethnic and Racial Studies*, 40(7): 1047–1057.

1 Exploring community and change from border perspective

> Traditionally, border studies have adopted a view from the center; we argue for a view from the periphery ... Rather than focusing on the rhetoric and intentions of central governments, we look at the social realities provoked by them.
>
> (Baud and van Schendel, 1997, p. 212)

The recent revival of studies on borders and boundaries has largely been influenced from the impact of globalization and proliferated in the academia with the concomitant questions of identity, culture, and space. A multi-disciplinary realm drawing on history, anthropology, sociology, political science, and geography today characterizes border studies. The erosion of the Cold War divide between the West and the Eastern Bloc with the collapse of the former Soviet Union and communist regimes at the turn of the 1990s and the accelerating globalization fostered interdisciplinary perspectives and ethnographic standpoints in border research (Paasi, 2011, p. 17). These processes and their analyses have contributed to the conflation of territorial and conceptual borders, the question of territorial sovereignty and identities, and cultural unity and difference within the same framework (Kolossov, 2005). The theories of globalization particularly stirred among the border scholars a critical geographical thinking and new spatial consciousness.

This book recognizes the valuable contribution of border studies in analyzing social stratification and change. In this chapter, I will review the scholarly literature on border studies with a critical eye towards delineating the main threads and conceptions of border as "peripheral" areas. A 'cultural turn' in border studies, which challenged the conventional notions of identity, culture, and territory, has questioned studies on border on the grounds that they reproduced the state-centric perspective with their classical themes such as cross-border cooperation and regional economic integration. Contemporary border studies might provide contributions to a more complex and enriched analysis of "state-society relationships," by problematizing a dualistic approach to the latter. In this chapter, two overriding theoretical approaches in border studies, namely liminality and borders-as-margins, are reviewed and singled out for their strengths and limitations for this study. Then, these threads are developed to show how the analysis of community and change could be pinpointed on the spatiality of border regions, embedded in networks of symbolic and economic exchanges. The last section is devoted to border studies in the Middle Eastern/Turkish context.

Borders as liminal spaces

Globalization theories revealed the fluidity of cultures, places, and identities as interstate borders became more permeable to the global flows, a term emphasized by globalization theorists such as Arjun Appadurai (1990) and Manuel Castells (1996). They addressed the shrinking of the globe with the enhanced global mobility of people, commodities, and ideas and the decentered nature of global processes. The arguments followed that the accelerated circulation of information, services, and ideas throughout the globe intensified the deterritorialization of state sovereignty, human diasporas, and even actual places. The globalization perspective revealed the emerging spaces of encounters and mobility, which are not nationally bounded and underlined the 'trans-national' context. According to Gupta and Ferguson, for instance, the more sophisticated communications and information network and better means of transporting goods and people created a transnational public sphere, which may forge new forms of solidarity and identity that are not spatially bounded (1992: 9). For these scholars, this did not point only to "the partial erosion of spatially bounded social worlds," but also to "the growing role of the imagination of places from a distance" (Gupta and Ferguson, 1992, p. 12). Thus, the rise of transnational spaces and communities was suggestive of the loss of a naturalized relationship between territory, nation, and identity, as well as multiple belongings embedded in more than one locality. These ideas shifted the focus of social analysis towards the fluid and multiple nature of identities.

The research on the Mexican–US border informed by globalization theories has promoted a "cultural turn" in border studies (Vila, 2003; Wilson and Donnan, 1998; Alvarez, 1995). I use here the cultural turn rather loosely to indicate these criticisms elaborated by Marcus and Clifford (1986) that promoted an anthropological writing similar to literary genres and situated the anthropologist's research experience as textually constructed meanings. A corollary to the cultural turn in social theory is the adoption of "border" as a category criticizing worn-out essentialist conceptions of culture, rooted in the Westphalian ideal of the nation-state.

Roberto Alvarez argues that the study of border or, as is it is often referred to as, borderlands constituted a new genre, "a basis upon which to redraw our conceptual frameworks of community and culture area" (Alvarez, 1995, p. 447). As Heyman and Campbell (2007b) argue, the inquiries of the Mexican–US migration and cultural flows had an exclusive contribution to this new genre. These studies became almost a model adopting the border as an image for the liminal space between two nation-states, in which nestles a mingling and multiplicity of identities. They enhanced the quest for hybrid cultures as accelerated flows and mobility crisscrossing the borderlands and endowed the border cultures with a subversive potential defying the ideologies of nation-state. In the new genre, boundaries acquire an analytical power in anthropological debates of culture and identity without any necessary reference to the geographical borders, which leads to an approach called by Alvarez (1995) "a-literalist" in contrast to its literal use.

The a-literalist approach has come under criticism on several grounds. Grimson (2006) and Vila (1997) point out to the adoption of research on the US–Mexican border almost as a paradigmatic model for border studies in general and a tendency to homogeneize borders (Grimson, 2006; Vila, 1997). Grimson argues that the paradigmatic status of the US–Mexican border gives way to a new ethnocentrism, while it also encourages the denaturalization of juridical borders and essentializes social identities. On the other hand, Vila emphasizes the employment of border-crossing as a metaphor for unravelling the question of border subjectivities by giving it privilege over the enclosure experience among border dwellers. Donnan and Wilson (1999) remind that the conception of borderlands has turned into an image that is not necessarily referring to a geographical border area and could be employed for the study of connections between cultures wherever these connections are found. These scholars denounce the tendency of this model to focus on social boundaries on geographical borders to the expense of hiding the latter out of sight.

These criticisms contend the borderlands genre by underlining the significance of geographical borders impinging on daily life in an age arguably towards a borderless world. This genre has emerged within the globalization context and emphasized the ideas of cross-border mobility and cultural hybridity. The criticisms contest it by putting emphasis on fragmentation and difference (Vila, 2000), conflict and growing inequality (Grimson, 2006), control and enclosure (Heyman and Cunningham, 2004), reterritorialization (van Schendel, 2005a), and spatial reproduction (Heyman and Campbell, 2009), immobility and entrapment (Navaro-Yashin, 2003) at the border zones. They propose a more balanced framework to reveal the interplay between geographical borders and social boundaries.

Criticisms also address the paradigmatic status of Mexican–US border anthropology, neglecting the relevance of national belongings and identifications in a global context. Vila's ethnographic account of the Mexican–US border illustrates that the border lies among the Mexican Americans living in diverse locations of the US as a constant reminder of their difference, giving to Mexicans living on the border a meaning of their identity as ethnicity and nationality simultaneously (Vila, 1997, p. 178). Similarly, the border geographer Anssi Paasi asserts that the territorial borders are central to the national ideological apparatus aiming to rejuvenate the state dominion over its citizens (Paasi, 2011, p. 21). Paasi, thus, urges the border researchers to pay attention to the role of practices and discourses that produce and reproduce the national borders in the discursive realm of state power. The 'boundaries are everywhere' thesis, as he calls it, omits the role of borders to strengthen the national community as a bounded unit (Paasi, 2011, p. 22). In short, his view underlines that the borders have become diffused rather than eroded and their changing meanings within the global context deserve much more attention.

The ethnographic studies reveal that material objects, people, and even funerals crossing the border can have symbolic value enforcing the distinction between members and outsiders (Reeves, 2007; Pelkmans, 2006; Parizot, 2008; Vila, 2000).

Unlike the above-reviewed a-literalist approach, these studies emphasize that border-crossing and increased interaction might actually reinforce social boundaries. It is understood that the a-literalist approach posits the sole border enforcer as the state as it privileges border-crossing and the transgression of cultural boundaries. Vila rightfully reminds of the evidence from research demonstrating that most Mexican Americans living in the US do support a closed border (Vila, 2000). To put it differently, the criticisms of the culturalist assumptions underline the bias that the transnational flows across the border breed the multiplicity of cultural belongings while they indicate to the function of territorial borders in giving rise to what Paasi (2011) calls border-producing identity narratives.

It is worth reminding that the themes and issues raised by the new borderlands genre are nevertheless relevant in broader border studies. The borderlands genre introduces to the border studies a constructivist view of transnationalism with a particular emphasis on the daily interactions of people making up the transnational spaces and communities. Moreover, it puts forth border as an analytical category to capture the transnational flows from below and emancipates the scholarly inquiry from the state-centric conceptions of territory, nation, and culture.

Borders as margins

I have mentioned in my review of the borderlands literature that this approach ascribes a subversive potential, which calls state sovereignty into question. It is true that the globalization process rendered border studies with a new impetus as more transnational flow of goods, people, and ideas tends to overcome international borders and erode state sovereignty. Yet, further studies of border warn against quick assumptions about the erosion of state sovereignty and argue instead in favor of analyzing more attentively the changing relationship between sovereignty and territory. A borders-as-margins approach posits the borders as state margins and views them as windows for the researcher to explore broader processes of state sovereignty, nationalism, and territorialization. Furthermore, while this approach problematizes the subversive capacity of border populations vis-a-vis state power, it embraces the knowledge of their transgressive experiences and practices at the state margins as indispensable for a critical inquiry of the state-imposed definitions of illegality.

Borders are often simultaneously geopolitical, socioeconomic, and cultural margins. They might be excluded from state resources or they could be marginalized by the shared culture of the nation because of the closer affinities across the border rather than with the broader society. Moreover, this approach adopts the research at the margins as an epistemological shift indicating "an analytical placement that makes evident both the constraining, oppressive quality of cultural exclusion and the creative potential of rearticulating, enlivening, and rearranging the very social categories that peripheralize a group's existence" (Tsing, 1994; cited in Galemba, 2013, p. 277). Border zones do not only constitute margins, but they are also marginalized. The border as state margins approach offers a framework that would not only challenge marginality discourses. It also demonstrates the ways in which these discourses could be assimilated by the state institutions as well as border dwellers to accommodate various interests (Obeid, 2010). These studies point to the cultural constructedness of borders as margins and the implications for social theory. For Veena Das and Deborah Poole, the ethnographic exploration of marginality and border construction offers "a unique perspective on the sorts of practices that seem to undo the state at its territorial and conceptual margins. Such margins are necessary entailment of the state" (Das and Poole, 2004, p. 4).

Janet Roitman's ethnography of illegal trade and gang-based road banditry at the Chad Basin, a geographical border junction among Nigeria, Niger, Camerun, Chad ve Central African Republic, employs a similar perspective to challenge the researcher's own analysis categories regarding the cross-border activities that do not conform to the state regulations of market economy. The adoption of state notions of illegality is a major impediment to understand how unregulated economic exchanges and financial relations emerge in the first place. The analysis of these activities deemed as criminal requires the widening of our scope beyond moralizing and stigmatizing definitions to capture their content. The ethnologist should recognize that "the transgression of physical frontiers (e.g. national borders) and conceptual boundaries (e.g. spoils becoming licit wealth) is critical to productive activity and the production of new forms of knowledge about the possibility of such activity" (Roitman, 2004a, pp. 16–17).

The illegalization of cross-border exchange practices by the law enforcers in fact contributes to the marginalization of poor border dwellers. Rather than its functions of providing security and order, it "enhances vulnerability; legitimizes exploitation; and justifies accumulation, extraction, and violence" (Galemba, 2013, p. 276). Yet the border dwellers might instrumentalize illegality in order to escape the more risky economic engagements and create alternative strategies of livelihood. Ethnography at the state margins may illustrate how the border dwellers negotiate their understanding of (il)legality (Baud and Schendel, 1997), navigate the boundaries between legal and illegal realms (Galemba, 2013), and the notions of legitimacy and an ethical conduct that partakes in their engagement in economic activities deemed illegal by the state (Galemba, 2008, 2012; Roitman, 2006). Heyman and Campbell direct their attention to the shifts and plays of discourses surrounding corruption among border actors and put forth cogently how they underline the border culture: The corruption at the Mexican–US border is regarded to be driven by a foreign culture, stigmatizing the Latino/Mexican populations even though the US agents engage in corrupt actions (Heyman and Campbell, 2007b, p. 208). They argue that the US border enforcers that are not familiar with the border culture mostly fall prey to such brokering arrangements.

Ethnography at the state margins may also contribute to our knowledge of state territorialization and sovereignty by focusing on the goods and people that weave in and out of legality. For instance, studies on post-colonial states of Asia demonstrate that the Westphalian notion of nation-state and its borders as impenetrable barriers was virtually non-existent until recently, as revealed by Wong's research on cross-border movements between Indonesia and Malaysia—transgressive crossings yet not stigmatizing Indonesians as illegal until 1963 (Wong, 2005). This is why Abraham and van Schendel define illegality as a "form of meaning produced as an *outcome* of the effect of a criminalized object moving between political, cultural, social and economic spaces" (Abraham and van Schendel, 2005, p. 16). Inquiring state efforts to criminalize the flows escaping its control reveals that the state is never able to monopolize regulatory practice (Baud and Schendel, 1997). These inquiries point to the competing regulatory authorities at border regions, which do not necessarily undermine state power (Abraham and van Schendel, 2005; Roitman, 2004b; 2004c).

Researchers may inquire into how state notions of illegality and border sur-veillance might contribute to our understanding about the fluidity of rebordering processes (Abraham and van Schendel, 2005) and the way these practices redress ethnic, class, gender, and sexual boundaries. The latter is mostly addressed in terms of a tension between state control and the implications of its evasion, which is played out in the language of cultural and gender identities. Cheater's (1998) article on sexual and gendered identities of Zimbabwean female traders crossing borders highlights the overlap between the patriarchal control over gender iden-tities and state efforts to control its border. As the economic crisis pushes most Zimbabwean women to opt for small-scale trade across the border as a livelihood strategy, this mobile population constitutes dangerous citizens for the state. Simi-larly Ana Alonso (2005) suggests that the spatial politics of security on the border are not only a state effect but also establish state sovereignty in gendered ways, discriminating between outlaw men and women, and unnaturalizing the latter.

These studies discuss the porosity of border and the transgression of boundaries as integral to the state functioning in a global system (van Schendel, 2005a; Gupta, 1995; Roitman, 2004a, 2007; Reeves, 2007; Das and Poole, 2004; Galemba, 2008). The border surveillance does not intend to effectively hold the frontier line against illegal intruders but to regulate these flows. Border enforcement is nec-essary for the perpetuity of state sovereignty and its territorial entity, rather than the obstruction of illegal entry of goods and people. The tight controls and strict regulations may be even more effectual in forging the conditions of illegality and informal economy (Andreas, 2000; cited in Galemba, 2008). The political scien-tist Peter Andreas reminds that the state monopolizes the power to criminalize and set the rules of the game by defining the content of illegal (Andreas, 2011).

This discussion is very much complemented by the debate regarding the illegal, informal, and small-scale economic activities, as well as forms of wealth erupting alongside the borders. Research pinpoints the importance of sub- and transna-tional regimes of accumulation as a critical connection among local economies, transnational flows, and mediating state institutions.

These studies focus on non-state channels utilized by border dwellers to resist the formal market sanctions and regulations, to manipulate, undermine, and undo the state regulatory authority rather than drawing on an ideal notion of moral economy and locating resistance as exogenous to the market and state. Roitman is particularly critical of the notion. She argues that the unregulated exchanges could enforce the state power even pluralized the regulatory authorities. According to Roitman (2004b), the distinction between state regulatory authority and state power is necessary. For example on Chad Basin, the authority figures evading the state are unemployed military disbanded from the mercenary, customs officers, big traders, and local chiefs linked to bureaucracy and politics; controlling the access to wealth and employment. These authority figures collected taxes at the customs or markets at the border regions, redemption prices and tribute money, protection fees, payments for safe delivery of goods, and established an autonomous fiscal base.

Thus, as these unregulated activities replace the state-led economic redistribution of wealth, it is misleading to call them informal or illegal. Roitman suggests that informal markets are erroneously regarded as emerging in nascent capitalism or "unproductive" economies, although they cannot be dissociated from dominant forms of production and distribution channels (Roitman, 1990). Carolyn Nordstrom asserts, drawing on her extensive research on war profiteering and illegal trade, that non-state relations of exchange and power weave across illegal, semi-legal, and informal markets at once (Nordstrom, 2000, p. 36). The fact that these activities are catalyzed through non-state channels does not mean that they weaken state power, but they co-exist as two different realms of authority and socio-economic organizations (Nordstrom, 2001, p. 219).

So, this perspective employs a conception of state as variously constructed by historically-situated and differentially-positioned actors (Donnan and Wilson, 1999, p. 106) and diffused relations of power (Roitman, 2004c). Roitman explicitly distinguishes between these unregulated activities and the moral economy as autonomous and dissenting relations flourished where the state power cannot establish its order (Roitman, 2006, p. 265). Such states of illegality rather constitute "a mode of establishing and authenticating the exercise of power over economic relations and forms of wealth, giving rise to political subjects who are at once subjected to governmental relations and active subjects within their realm" (Roitman, 2006, p. 264). They problematize the "subversive" nature of illegal practices and exchanges and urge to carefully delineate the limits of anti-state resistance on the basis of rigorous investigation (Galemba, 2013).

In sum, the "borders-as-margins" approach draws on the post-modern theories of power—such as biopolitics and affect—in order to inquire into the micro-realities of the state–society relations. Study of border dwellers' everyday struggles to manipulate and circumvent the border offers insight to the state actions to impose its control and regulation over them and contributes to our understanding of translocal processes through which the state is experienced (Wilson and Donnan, 1998; Gupta, 1995; Das, 2004). Veena Das conceptualizes the state margins in both a territorial and a conceptual sense to analyze how the state is experienced and undone through its illegibility. Her viewpoint inspires thinking of border regions as state margins, where the heavy state apparatus with its border customs and military control, oscillating between being vigilant and complicitious, makes the state utterly illegible for the border dwellers and therefore manipulated.

A major criticism to the borders as margin approach is its analytic bias that forces the actions of state power as well as border dwellers' to fit into a

domination-resistance scheme. Heyman and Campbell (2007a) oppose this analytical framework by emphasizing the oblique actions of human individuals indi-
rectly confronting and evading the state power and its institutions. Drawing on an
ethnography on ramshackle, mobile, and uneasy to discern settlements of Mexican
migrants near to the US–Mexican border, called *colonias*, the authors argue that the
relationship of colonia dwellers with the US population census agents, challenged
by the complex and provisional nature of the settlement forms and zigzag behaviour
of migrations back-and-forth resist their placement into a domination-resistance
framework. The reluctance of colonia settlers to cooperate with the US agents, their
limited competency in English, the hard-to-discern nature of the dwellings, and
continually shifting dwelling environment result in difficulties in estimating the
accurate population census among Mexican migrants living in the US.

Although the migrants engage in illegal alternatives of labor employment and
border-crossing defying the dominant actors—that is, state regulations—their acts
do not generally consist of intentional claims of resistance, but simply fear and
distrust for the US law enforcement and survival efforts in a new social context
that they are not much cognizant of. Hence, the authors advocate for an analytical framework to capture the oblique, improvised, and mobile actions indi-
rectly related to power situations, which they call slantwise, and to handle these
behaviours and actions with greater subtlety within a framework that hinges upon
both resistance and domination. To put it differently, they do not deny that the
slantwise actions of migrants could be orchestrated in resistance or internalization
of domination, but these would be consequences or effects, rather than explanatory causes.

To conclude, I reviewed two approaches in this section, namely, the borders
as liminality and as margins that give the concept of border an analytic priority
to develop a research agenda on questions of identity, citizenship, territoriality,
and state power. These approaches add to our understanding of national, ethnic,
gender, and sexual identities since borders alter these identities in ways not found
elsewhere. They provide insights into the working of state power at the level of
everyday life, as well as to the practices that challenge and undo it. Though these
two approaches oscillate between a perspective that views state power either as
eroded or enforced at the border regions, they both point to the functions of border control and surveillance operating not only at the politico-juridical level of
regulating population flows but also at the level of meaning construction (Wilson
and Donnan, 1998). In brief, border studies focus on mobilities and enclosures in
a globalized world to explore borders not only as political and economic barriers
and prospects, but also as social and moral systems.

Resituating the study of borderlands

The above-reviewed perspectives in border studies may offer an innovative
approach to the conventional analysis about social stratification and change in the
borderlands, but their limitations should also be noted. The constructivist view of
the borders offers a perspective for symbolic and cultural processes involved in the

reproduction of borders. They provide tools to conceive how they are reproduced at the center rather than peripheral border zones and by the state's efforts through a set of practices and discourses. By underlining the role of identities, symbols, and meanings circulating through transnational flows, they contribute to our understanding of how borders are "rebordered and debordered" (Staudt and Spener, 1998, quoted in Heyman and Cunningham, 2004, p. 292). Yet, as it focuses on the symbolic reproduction, it de-emphasizes the "real" relations between individuals, producing borders as social spaces. On the other hand, the view of borders as state margins provides a framework for the multiple configuration of social dynamics at the border impinging on the heart of states, but it tends to simplify it on the basis of a dichotomy between internalization of domination and resistance to it.

As van Schendel argues, the historicity of border spaces is essential to understand how social relations at borders are continually being reconfigured (van Schendel, 2005b, p. 9). Globalization creates new spaces and territorialities through transnational flows that the states have to deal with at sub-national and supra-national scales. The increasing prominence of cross-border cooperation, regional economic integration, and new border management with supra-national institutions illustrate a politics of scale, as van Schendel (2005b) says, that the states are engaging in order to reassert their territorial sovereignty. Thus, analyses of globalization as de-territorialization and re-territorialization require an understanding of the ways in which transnational flows are mediated by the states. The states are compelled to re-design their technologies of border surveillance as in the case with European Union (EU)-led Frontex, set new schemes of immigration and naturalization and facilitate international trade by appealing to multilateral bodies and arrangements. State powers in a globalizing context wobble between tighter control and increasing permeability at the borders. Van Schendel argues that these policies have significant repercussions at the border regions, by turning them into "spaces of engagement" where various domains of power and profit are overlapped. Thus, transborder arrangements highlight the function of border societies to ease or impede them.

The historicity of border spaces can be revealed by inquiring how they "act as pivots between territorial states and transnational flows (as well as between separate flows)" (van Schendel, 2005a, p. 62). Border scholars emphasize that border communities are not passive recipients of such flows but they are part of a social system straddling the border that regulate and restructure these flows (van Schendel, 2005b; Alvarez, 1995). The mere emphasis on state governance and transborder flows tends to obscure the border communities and their incorporation among the particular dynamics influential at border regions. It is this ongoing, dialectical process among border communities, territorial states, and transnational flows that give distinctive characteristics to a border place on an international boundary. This viewpoint undermines the simplistic counterposing border communities vis-a-vis the state.

Borders do not only separate social groups but they also tempt them to interact thanks to spatial proximity and social affinity. This is one major reason why the interstate conflicts and tensions affect the border communities most. Because a "historical consociationality" (Wong, 2005) often governs the cross-border movements rather than a border logic, border communities can appropriate the territorial borders for their own political agendas and as Sahlins' study of the Franco-Spanish border at Cerdanya valley suggests, they can play one state off against other (Sahlins, 1998). The contingent and unintended nature of transborder arrangements that mediate between the state politics of scale and transnational flows turn the border regions into contentious zones with unique social dynamics and historical development.

Borders are not only clusters of cross-border ties that evolve across space but also time. Alvarez puts particular emphasis on the "specific permanence and longevity of border people in forming lasting social bonds and in political economic struggle," a disregarded fact by most studies of border (Alvarez, 1995, p. 462). Cross-border networks and ethnic ties may sustain themselves even after the imposition of new territorial boundaries, a condition renowned for J.W. Cole and Eric Wolf's study of Italian Tyrol as *The Hidden Frontier* (1974). The authors describe how the ethnic and nationalist loyalties could be strongly maintained in two villages in north Italy that was initially part of Australian Tyrol before its transference to Italy in 1919. On the other hand, the delineation of new geopolitical borders may also instigate the emergence of new networks and confrontations across border communities.

Similarly, geographer David Newman (2003) underlines the processes through which borders are demarcated and delineated for a historical perspective to the border dynamics and nature of cross-border networks. These processes are likely to define the management and degree of permeability of the border. Territorial boundaries generally do not match the imagined territorialities of border communities and this is precisely why, as Newman argues, border dynamics rise and contest. The demarcation of boundary lines involves not only geographic mapping and plotting but it sets the parameters for demarcation. It draws on the state politics of territoriality, its rules and regulations about inclusion/exclusion, and their instantiation as cultural differences. Newman reminds that the demarcation of a border may alter the power settings among the border community enclosed by it (Newman, 2003, p. 22). Border dwellers may benefit or lose unequally from being enclosed. Law enforcement, local elites, and lower-strata border people may be aligned differently in relation to the bordering process or depending on their participation/exclusion from the decision-making. As illustrated by the colonial settings, the demarcation process may shift the power configurations among state, local elite, and border people based on collaboration or confrontation with colonial powers.

Finally, border towns and cities are proper settings to observe the changing configurations of power relations and to inquire into the historicity of border spaces. Paul Nugent (2012) in a comparative study of border towns claims that these places provide "a telling insight into the geographies of wealth and power" (Nugent, 2012, p. 557). As Nugent suggests, border towns offer an understanding into the ways in which the new and hybridized form of governance at the border oscillates between tighter control and increased permeability. Urban settings at the border may come out in a diverse range: as industrial–military plants like the small-arms factories of the Pakistan–Afghanistan frontier and maquiladora plants of the US–Mexican frontier; as casinos and heroin refineries on the Thai–Burmese frontier; as smugglers' districts on Benin–Nigeria or Spanish–Morocco frontiers; as prostitution brothels at Czech–German and US–Mexican frontiers or as refugee enclaves in a number of borders (Abraham and van Schendel, 2005). Border towns may act as hubs or entrepots for the re-routing of international and clandestine trade, as is particularly highlighted by the African context (Nugent, 2012; Roitman, 2004c). They make visible the networking around the border necessary for transborder arrangements.

Border towns can also make visible the historical trajectories of urbanization in the periphery or emergence of new urban forms at the center or at their outskirts. Geographies of the US–Mexican border are exemplary with the emerging districts of prostitution as well as shanties known as *colonias*. Curtis and Arreola (1991) not only survey the spatial effects of new economies at the border creating prostitution districts but they also underline the iconic images of border towns due to the flowing population, goods, and signs swirling around them. For example, the naming of the red-light districts as zones of tolerance or, in short, *zonas* illustrate how far unsavory adult entertainment is associated with these border towns, though they are not more common than in any large Mexican city.

Border towns may accentuate various lifestyles, representations, and policies in a clash with each other. Pelkmans' (2006) ethnography of Batumi in Georgia inquires about the ways in which the flows of new images, ideals, and goods circulated into the city after the fall of the iron curtain with Turkey and the opening of the border collides with the ongoing nepotism and state control inherited from the former Soviet rule. He dedicates a chapter to a number of state construction projects that accelerated in the late 1990s in order to provide insight into the uneasy sociopolitical change. To note another example, the urban settings at the US–Mexican border are spatial embodiments of crime and impoverishment in the eyes of the larger American public. Critical ethnography, on the other hand, reveals the uneven development carving out these spaces, turning the farmers into real estate entrepreneurs that capitalize on the nonexistent zoning laws in colonias (Hill, 2003). Therefore, historical surveys of the border towns may direct the researcher's attention to the inner contradictions and tensions of community-building and the social reproduction of class positions in changing border settings. Border towns point not only to slantwise perspectives and actions that the subaltern engage in for improvising, adapting, seeking social mobility but also to the elite's behaviours.

Studying borders in Turkey and the Middle East: the challenge of limited research

The Middle Eastern context is not much informed by the novel approaches in border studies. The Middle Eastern borders are usually associated with stereotypical assumptions.[1] It is assumed that Middle Eastern borders are artificially drawn by colonial powers and they are characterized by kinship and ethnic affinities and illegal transactions among border communities. Scholarly literature also tends to identify these borders by the vast discrepancy between territorial borders and social boundaries (Brandell, 2006) and the illegal transactions such as contraband, which arise from the mismatch between borders and the economic domains they delineate (Nordstrom, 2011). Territorial and social boundaries of the Middle East are not subject to rigorous investigation and do not attract as much interest as the well-established schools of African and European border studies or the developing research tradition on Asian borders. The achievements of scholarship on the latter three regions, as well as on the US–Mexican border, benefit from the advantages of comparative studies (Baud and van Schendel, 1997; Nugent, 2012; Asiwaju, 1994), whereas it is not possible to draw comparisons with the Middle East.

The same goes for Turkish borders. Limited research will not allow me to review, but I will at least draw affinities and divergences between the scholarly literature about the Turkish context and broader border studies. Earlier studies on Turkish borders have been influenced from the state-centered perspective, underlying the state's politics of nationalism with particular emphasis on the Kurdish ethnic question and a strategist approach (Özgen, 2004). They highlighted the role of state borders and state rhetorics on shaping the boundaries of citizenship. Studies concerning the Kurdish-populated border geographies have been largely framed by debates of the Asian mode of production, dependency school, and world system theory developed on the basis of the center-periphery model (Akyüz, 2013, p. 109).

Recent studies draw on ethnographic research and archival resources, as well as discourse analysis. I have asserted that the conception of borders as liminal spaces encourages transnationalism analyses on the basis of border-crossings and cultural hybridity. The Turkish context on the contrary highlights nation as the main mode of identification at the border, though multiple legal and cultural belongings may emerge to interact with it. Against the globalization theories that underline the importance of global flows and culture over the nation-state, I refer here to the scholars arguing that nations and their boundaries continue to draw distinctions about inclusion and exclusion, about members and outsiders (Dragojlovic, 2008). Transnationalism in the Turkish context is much concerned with the tensions and contradictions in relation to citizenship and exclusion questions (Cizre, 2001; İçduygu and Kaygusuz, 2004).

Studies about Turkish borders focus on the border regime and bordering processes in a historical perspective (Kaşlı and Parla, 2009; Kaşlı, 2014; Özgen, 2006, 2007a, 2007b; Beller-Hann and Hann, 1998; Pelkmans, 2006; Akyüz, 2013, 2014; Genç, 2015). There is also a limited but promising realm of investigation with rigorous historical research on the Greek–Ottoman border (Gavrilis, 2008), Ottoman–Iranian border (Ateş, 2011), and Turkish–Syrian border (Altuğ and White, 2009). There are exceptions to the overriding debates of identity with reference to transnationalism, Hatay at the Turkish–Syrian border being the typical research for placing arguments of triumphant cultural hybridity and multiculturalism under the impact of globalization (Stokes, 1998; Doğruel, 2013). Transnationalism is also investigated in terms of 'globalization from below', i.e. everyday interactions and networking on the ground among ordinary people in a study on undocumented or illegal transit migration (İçduygu and Toktaş, 2002) and on the shuttle trade between Russia and Turkey (Yükseker, 2007).

The transnationalism debate in the Turkish context prioritized the boundary processes redefining national and cultural belongings as well as the shifting meanings of territorial boundaries, turning the scale for cultural analyses though attentive to the fragmentation, difference, and new forms of inequalities created by the globalization context. Among these studies, research on geographic border settings that are insightful about state sovereignty, nation-building, and the politics of territorialization are far from satisfying but it is a promising sub-area clinging to the borders-as-margins approach. In addition to Kaşlı and Parla (2009) and Özgen (2007a) cited earlier, more recent border ethnographies of transboundary informal economies and geographical mobility point to the emergence of new territorialities and sovereignty practices as in the case of Kurdish communities at the triple borders of Iran, Iraq, and Turkey (Özcan, 2014) and Syrian Kurdistan (Yıldız, 2014) indicating to the fragmented nature of territorialities and sovereignties that may be superimposed on each other.

Nonetheless, a major incentive in proliferating border studies within the Turkish context seemingly will be the global crisis in the refugee reception, triggered by the post-2011 Syrian conflict and the new movements of migrations to Europe. The refugee movements highlighted Turkey as a neighboring country, which became destination and transit after the Syrian conflict, while attracting attention to its shifting border regime and emergent border politics (Heck and Hess, 2017; Baban et al., 2017; Kaşlı, 2017). A snapshot of the impact of the Syrian conflict on Kilis borderlands draws widely on these references in the Epilogue. Hence, it suffices here to mention that these studies promise new contributions to the transnationalism debate in the Turkish context as they demonstrate how the Turkish policies of migration and border control as well as refugee subjectivities are shaped amidst the transnational political and juridical processes, a humanitarian discourse appealing to the security concerns and contested forms of refugeeization.

Concluding remarks

This chapter provides a critical review of the scholarly literature on border studies and goes over the main themes and debates, suggesting the analytical use and significance of border and boundary concepts for social theory. Border studies have made important contributions to globalization and identity theories, but their utmost achievement is to make the idea that geographical borders are a proper object of research rather than a mere backdrop widely accepted.

Border studies suggest an anthropological perspective from the periphery against the state-centered analysis. I distinguish the two theoretical approaches in contemporary border studies. The liminality approach utilizes a cultural study of borders and boundaries and conceptualizes the borderlands as spaces for mingling and multiplicity of cultural identities. The criticisms addressed to the liminality approach for its a-literalism and overemphasis on border-crossing as a metaphor for identity formation adjust this approach to restore its operationality for border studies. The contribution of the liminality approach for this study is to provide a research framework that conflates territorial and conceptual boundaries. The borders-as-margins approach is primarily focused on the territorial borders as research sites and regards them as windows opening to the broader studies of nation-building, citizenship, state sovereignty, and territoriality. The contribution of the borders-as-margins approach for this study is not only to introduce an epistemological shift from the state-centered perspective towards the periphery, but also to reveal the power mechanisms that peripheralize a group's existence.

Nevertheless, both approaches suffer from a resistance-internalized domination dichotomy, rather than placing border studies in a continuum that hinges upon both resistance and internalized domination. The borders as geographical, socio-economic, and cultural margins do not have to fit necessarily within this dichotomy, as most of the behaviours and actions of border dwellers are oblique and improvised without the intention to oppose state ideologies or challenge state regulations and control. Border dwellers rather navigate, manipulate, and evade the territorial and conceptual (legal, social) boundaries. Hence, this literature review redresses the study of the borderland and its dynamics in order to understand the inner stratification and change in Kilis town.

Note

1 I believe that such views reify Middle Eastern borders and partially recognize the truth. These are not necessarily the sole characteristics of the Middle East, nor do these disparities among political, social, and economic domains only exist there. For instance, strong kinship ties, arguably typical in the Middle East, are even more dominant in Central Asia where Ferghana Valley delineates Uzbekistan–Kyrgyzstan–Tajikistan borders—a fact exposed by the very recent border closures in 1999 in this post-Soviet geography (Megoran, 2006; Reeves, 2007).

References

Abraham, I. and van Schendel, W (2005) "Introduction: The Making of Illicitness", pp. 1–37 in W. van Schendel and I. Abraham (eds.), *Illicit Flows and Criminal Things: States, Borders, and the Other Side of Globalization*, Bloomington: Indiana University Press.

Akyüz, L. (2013) *Ethnicity and Gender Dynamics of Living in Borderlands: The Case of Hopa-Turkey*, unpublished PhD dissertation, METU, Ankara.

Akyüz, L. (2014) "Liminal Alanlar olarak Sınırlar: Türkiye-Gürcistan Sınırında Ekonomik Yaşam ve Etnik Kimliklerin Sınır Deneyimleri", *Toplum ve Bilim*, 131: 84–103.

Alonso, A. M. (2005) "Sovereignty, the Spatial Politics of Security, and Gender: Looking North and South from the US–Mexico Border", pp. 27–52 in C. Krohn-Hansen and K. G. Nustad (eds.), *State Formation: Anthropological Perspectives*, London; Ann Arbor, MI: Pluto Press.

Altuğ, S. and White, B. T. (2009) "Frontières et pouvoir d'État. La frontière turco-syrienne dans les années 1920 et 1930", *Vingtième siècle*, 103: 91–104.

Alvarez, R. (1995) "The Mexico–U.S. Border: The Making of an Anthropology of Borderlands", *Annual Review of Anthropology*, 24: 447–470.

Andreas, P. (2000) *Border Games: Policing the US–Mexico Divide*, Ithaca: Cornell University Press.

Andreas, P. (2011) "Illicit Globalization: Myths, Misconceptions, and Historical Lessons", *Political Science Quarterly*, 126(3): 403–425.

Appadurai, A. (1990) "Disjuncture and Difference in the Global Culture Economy", *Theory, Culture, and Society*, 7: 295–310.

Asiwaju A. I. (1994) "Borders and Borderlands as Linchpins for Regional Integration in Africa: Lessons of the European Experience", pp. 57–75 in C. H. Schofield (ed.), *World Boundaries Vol. 1 (Global Boundaries)*, London: Routledge.

Ateş, S. (2011) "Bones of Contention: Corpse Traffic and Ottoman-Iranian Rivalry in Nineteenth-Century Iraq", *Comparative Studies of South Asia, Africa and the Middle East*, 30(3): 521–532.

Baban, F., Ilcan, S. and Rygiel, K. (2017) "Playing Border Politics with Urban Syrian Refugees: Legal Ambiguities, Insecurities, and Humanitarian Assistance in Turkey", *Movements: Journal for Critical Migration and Border Regime Studies*, 3(2): 79–100.

Baud, M. and van Schendel, W. (1997) "Toward a Comparative History of Borderlands", *Journal of World History*, 8(2): 211–242.

Beller-Hann,I. and Hann, C. (1998) "Markets, Modernity and Morality in North-East Turkey", pp. 237–262 in T. Wilson and H. Donnan (eds.), Border Identities: Nation and State at International Frontiers, Cambridge: Cambridge University Press.

Brandell, I. (2006) "Introduction", I. Brandell (ed.), *State Frontiers: Borders and Boundaries in the Middle East*, London, New York: I.B. Tauris.

Castells, M. (1996) *The Rise of the Network Society, The Information Age: Economy, Society and Culture*, Vol. I. Cambridge, MA; Oxford: Blackwell.

Cheater, A. P. (1998) "Transcending the State? Gender and Borderline Constructions of Citizenship in Zimbabwe", pp. 191–214 in T. Wilson and D. Hastings (eds.), *Border Identities: Nation and State at International Frontiers*, Cambridge: Cambridge University Press.

Cizre, Ü. (2001) "Turkey's Kurdish Problem: Borders, Identity and Hegemony", pp. 222–250 in B. O'Leary, I. S. Lustick and T. Callaghy (eds.), *Rightsizing the State: The Politics of Moving Borders*, Oxford; New York: Oxford University Press.

Cole, J. W. and Wolf, E. R. (1974) *The Hidden Frontier: Ecology and Ethnicity in an Alpine Valley*, New York: Academic Press.

Curtis J. R. and Arreola, D. D. (1991) "Zonas de Tolerancia on the Northern Mexican Border", *Geographical Review*, 81(3): 333–346.

Das, V. and Poole, D. (2004) "State and Its Margin: Comparative Ethnographies", pp. 3–34 in V. Das and D. Poole (eds.), *Anthropology in the Margins of the State*, Santa Fe: School of American Research Press.

Doğruel, F. (2013) "An Authentic Experience of 'Multiculturalism' at the Border City of Antakya", *Çağdaş Türkiye Tarihi Araştırmaları Dergisi/Journal of Modern Turkish History Studies* XIII/26 (Bahar/Spring), 273–295.

Donnan, H. and Wilson, T. M. (1999) *Borders: Frontiers of Identity, Nation and State*, Oxford: Berg.

Dragojlovic, A. (2008) "Reframing the Nation: Migration, Borders and Belonging", *The Asia Pacific Journal of Anthropology*, 9(4): 279–284.

Galemba R. (2008) "Informal and Illicit Entrepreneurs: Fighting for a Place in the Neoliberal Economic Order", *Anthropology of Work Review*, 29(2): 19–25.

Galemba, R. (2012) "'Corn is Food, Not Contraband': The Right to 'Free Trade' at the Mexico–Guatemala Border", *American Ethnologist*, 39(4): 716–734.

Galemba, R. (2013) "Illegality and Invisibility at Margins and Borders", *Political and Legal Anthropology Review*, 36(2): 274–285.

Gavrilis, G. (2008) "The Greek-Ottoman Boundary as Institution, Locality, and Process 1832–1882", *American Behavioral Scientist*, 51(10): 1512–1537.

Genç, D. (2015) "An Analysis of Turkey's Bordering Processes: Why and Against Whom?", *Turkish Studies*, 16(4): 527–533.

Grimson, A. (2006) "Cultures are More Hybrid than Identifications: A Dialogue on Borders from the Southern Cone", *Latino Studies*, 4: 96–119.

Gupta, A. (1995) "Blurred Boundaries: The Discourse of Corruption, the Culture of Politics, and the Imagined State", *American Ethnologist*, 22(2): 375–402.

Gupta, A. and Ferguson, J. (1992) "Beyond 'Culture': Space, Identity, and the Politics of Difference", *Cultural Anthropology*, 7(1): 6–23.

Heck, G. and Hess, S. (2017) "Tracing the Effects of the EU–Turkey Deal: The Momentum of the Multi-layered Turkish Border Regime", *Movements: Journal for Critical Migration and Border Regime Studies*, 3(2): 35–56.

Heyman, J. and Campbell, H. (2007a) "Slantwise: Beyond Domination and Resistance on the Border", *Journal of Contemporary Ethnography*, 36: 3–30.

Heyman, J. and Campbell, H. (2007b) "Corruption in the U.S. Borderlands with Mexico: The 'Purity' of Society and the 'Perversity' of Borders", pp. 191–217 in M. Nuijten and G.Anders, (eds.), *Corruption and the Secret of Law: A Legal Anthropological Perspective*, Aldershot: Ashgate.

Heyman, J. and Campbell, H. (2009) "The Anthropology of Global Flows: A Critical Reading of Appadurai's 'Disjuncture and Difference in the Global Cultural Economy'", *Anthropological Theory*, 9(2): 131–148.

Heyman, J. and Cunningham, H. (2004) "Introduction: Mobilities and Enclosures at Borders", *Identities: Global Studies in Culture and Power*, 11: 289–302.

Hill, S. (2003) "Metaphoric Enrichment and Material Poverty: The Making of 'Colonias'", pp. 141–167 in P. Vila (ed.), *Ethnography at the Border*, Minneapolis: University of Minneapolis Press.

İçduygu, A. and Kaygusuz, Ö. (2004) "The Politics of Citizenship by Drawing Borders: Foreign Policy and the Construction of National Citizenship Identity in Turkey", *Middle Eastern Studies*, 40(6): 26–50.

İçduygu, A. and Toktaş, Ş. (2002) "How Do Smuggling and Trafficking Operate via Irregular Border Crossings in the Middle East?", *International Migration*, 40(6): 25–52.

Kaşlı, Z. (2014) "Kimine duvar kimine komşu kapısı: Türkiye-Yunanistan sınırının seçici geçirgenliği", *Toplum ve Bilim*, 131: 44–68.

Kaşlı, Z. (2017) "A Tale of Two Cities: Multiple Practices of Bordering and Degrees of 'Transit' in and through Turkey", *Journal of Refugee Studies*, 29(4): 528–548.

Kaşlı, Z. and Parla, A. (2009) "Broken Lines of Il/Legality and the Reproduction of State Sovereignty: The Impact of Visa Policies on Immigrants to Turkey from Bulgaria", *Alternatives*, 34(2): 203–227.

Kolossov, V. (2005) "Border Studies: Changing Perspectives and Theoretical Approaches", *Geopolitics*, 10(4): 606–632.

Marcus, E. G. and Clifford, J. (1986) *Writing Culture: The Poetics and Politics of Ethnography*, Berkeley: University of California Press.

Megoran, N. (2006) "For Ethnography in Political Geography: Experiencing and Re-Imagining Ferghana Valley Boundary Closures", *Political Geography*, 25: 622–640.

Navaro-Yashin, Y. (2003) "'Life is Dead Here': Sensing the Political in 'No Man's Land'", *Anthropological Theory*, 3(1): 107–125.

Newman, D. (2003) "On Borders and Power: A Theoretical Framework", *Journal of Borderland Studies*, 18(1): 13–25.

Nordstrom, C. (2000) "Shadows and Sovereigns", *Theory, Culture and Society*, 17(4): 35–54.

Nordstrom, C. (2001) "Out of Shadows", pp. 216–239 in M. Barnett et al. (eds.), *Intervention and Transnationalism in Africa: Global-Local Networks of Power*, Cambridge: Cambridge University Press.

Nordstrom, C. (2011) "'Extra-Legality in the Middle', Illicit Crossings: Smuggling, Migration, Contraband", *Middle East Report* 261(41): 10–13.

Nugent, P. (2012) "Border Towns and Cities in Comparative Perspective: Barriers, Flows and Governance", pp. 557–572 in T. Wilson and H. Donnan (eds.), *A Companion to Border Studies (Blackwell Companion to Anthropology)*, Oxford: Blackwell.

Obeid, M. (2010) "Searching for the 'Ideal Face of the State' in a Lebanese Border Town", *Journal of the Royal Anthropological Institute*, 16(2): 330–346.

Özcan, Ö. (2014) "Yüksekova'da Sınır Deneyimleri: Bir "Sınır Kaçakçılığı" Hikâyesi ve Barış Süreci", *Toplum ve Bilim*, 131: 162–185.

Özgen, N. (2004) "Sınır Kasabaları Sosyolojisi", unpublished report for research project sponsored by TUBITAK and Ege Üniversity Bilim-Teknoloji Uygulama ve Araştırma Merkezi (EBILTEM), İzmir. Retrieved 13 September 2014 from http://neseozgen.net/wp-content/uploads/26.pdf.

Özgen, N. (2006) "Vatanın Çerçevesi Olarak Sınırlar: Haco Ağa ve Alakamış Katliamı Örneği" *Tiroj*, Eylül-Ekim.

Özgen, N. (2007a) "Devlet, Sınır, Aşiret: Aşiretin Etnik Bir Kimlik Olarak Yeniden İnşası", *Toplum ve Bilim*, 108: 239–261.

Özgen, N. (2007b) "Öteki'nin Kadını: Beden ve Milliyetçi Politikalar", *Türkiye'de Feminist Yaklaşımlar*, no. 2. Retrieved 13 September 2014 from www.feministyaklasimlar.org/sayi-02-subat-2007/otekinin-kadini-beden-ve-milliyetci-politikalar/.

Paasi, A. (2011) "A 'Border Theory': An Unattainable Dream or a Realistic Aim for Border Scholars?" pp. 11–31 in D. Wastl-Walter (ed.), *A Research Companion to Border Studies*, Aldershot: Ashgate.

Parizot, C. (2008) "Crossing Borders, Retaining Boundaries: Kin-nections of Negev Bedouin in Gaza, West Bank, and Jordan", pp. 58–84 in S. Hanafi (ed.), *Crossing Borders, Shifting Boundaries: Palestinian Dilemmas*, Cairo: American University in Cairo Press.

Pelkmans, M. (2006) *Defending the Border: Identity, Religion, and Modernity in the Republic of Georgia (Culture and Society after Socialism)*, Ithaca, NY: University of Cornell Press.

Reeves, M. (2007) "Unstable Objects: Corpses, Checkpoints and 'Chessboard Borders in the Ferghana Valley", *Anthropology of East Europe Review*, 25(1): 72–84.

Roitman, J. (1990) "The Politics of Informal Markets in Sub-Saharan Africa", *The Journal of Modern African Studies*, 28(4): 671–696.

Roitman, J. (2004a) "Introduction", pp. 1–22 in *Fiscal Disobedience: An Anthropology of Economic Regulation in Central Africa*, Princeton, NJ: Princeton University Press.

Roitman, J. (2004b) "Productivity in the Margins: The Reconstitution of State Power in the Chad Basin", pp. 191–224 in D. Poole and V. Das (eds.), *Anthropology at the Margins of the State*, Santa Fe: School of American Research Press.

Roitman, J. (2004c) "The Garrison-Entrepôt: A Mode of Governing in the Chad Basin", pp. 417–436 in A. Ong and S. J. Collier (eds.), *Global Assemblages: Technology, Politics, and Ethics as Anthropological Problems*, Hoboken, NJ: Wiley-Blackwell.

Roitman, J. (2006) "The Ethics of Illegality in the Chad Basin", pp. 247–272 in J. Comaroff and J. Comaroff (eds.), *Law and Disorder in the Postcolony*, Chicago: University of Chicago Press.

Roitman, J. (2007) "The Right to Tax: Economic Citizenship in the Chad Basin", *Citizenship Studies*, 11(2): 187–209.

Sahlins, P. (1998) "State Formation and National Identity in the Catalan Borderlands during the Eighteen and Nineteenth Centuries", pp. 31–61 in T. M. Wilson and H. Donnan (eds.), *Border Identities: Nation and State at International Frontiers*, Cambridge: Cambridge University Press.

Staudt, K. and David, S. (1998) "The View from the Frontier: Theoretical Perspectives Undisciplined", pp. 3–34 in D. Spener and K. Staudt (eds.), *The U.S.–Mexico Border: Transcending Divisions, Contesting Identities*, Boulder, CO: Lynne Rienner.

Stokes, M. (1998) "Imagining 'the South': Hybridity, Heterotopia and Arabesk on the Turkish–Syrian Border", pp. 263–289 in T. Wilson and H. Donnan (eds.), *Border Identities: Nation and State at International Frontiers*, Cambridge: Cambridge University Press.

Tsing, A. (1994) "From the Margins", *Cultural Anthropology*, 9(3): 279–297.

van Schendel, W. (2005a) "Spaces of Engagement: How Borderlands, Illicit Flows and Territorial States Interlock", pp. 38–68 in I. Abraham and W. van Schendel (eds.), *Illicit Flows and Criminal Things: States, Borders, and the Other Side of Globalization*, Bloomington: Indiana University Press.

van Schendel, W. (2005b) "Studying Borderlands", pp. 1–24 in *The Bengal Borderlands: Beyond State and Nation in South Asia*, London: Anthem Press.

Vila, Pablo (1997) "Narrative Identities: The Employment of the Mexican on the U.S.–Mexican Border", *Sociological Quarterly*, 38(1): 147–194.

Vila, Pablo (2000) *Crossing Borders, Reinforcing Borders: Social Categories, Metaphors, and Narrative Identities on the U.S.–Mexico Frontier*, Austin: University of Texas Press.

Vila, Pablo (2003) "Introduction: Border Ethnographies", pp. ix–xxxv in P. Vila (ed.), *Ethnography at the Border (Cultural Studies of the Americas)*, Minneapolis: University of Minnesota.

Wilson, T. and Donnan, H. (1998) *Border Identities: Nation and State at International Frontiers*, Cambridge: Cambridge University Press.

Wong, D. (2005) "The Rumor of Trafficking: Border Controls, Illegal Migration, and the Sovereignty of the Nation-State", pp. 69–100 in W. van Schendel and I. Abraham (eds.), *Illicit Flows and Criminal Things: States, Borders, and the Other Side of Globalization*, Bloomington: Indiana University Press.

Yıldız, E. (2014) "Kaçak Pazarlar, Tutuk(lu) Hareketlilikler: Antep'in İran Pazarı, Kaçağın Emeği ve Sınırötesilikleri Üzerine bir Deneme", *Toplum ve Bilim*, 131: 186–207.

Yükseker, D. (2007) "Shuttling Goods, Weaving Consumer Tastes: Informal Trade between Turkey and Russia", *International Journal of Urban and Regional Research*, 31(1): 60–72.

Part I
Traditional notables

2 *Eşraf* and *esnaf*

Local notables and drawing social boundaries

Kilis town cannot be compared to the spectacular ancient city of Aleppo, which was covered by the United Nations Educational, Scientific and Cultural Organization's (UNESCO) list of world heritage sites—though today it is wretched because of the war in Syria. But the old town center still retains imprints of regional architectural texture with stone monuments, narrow streets, stone-laid barrel-vaulted passages, and blind alleys. The urban sprawl has engulfed the old town and the derelict buildings by recycling them into the urban development, making the ancient architecture (except the restored monumental buildings) appear ramshackle. Until recently, the town landscape was a testimony to the lineage and patrimony of local notables rooted in the late Ottoman era. Presently, it seems to attest to notables' fall from grace. This chapter explores the notables' experience of falling from grace at the southeastern margins of Turkey and its repercussions on their sense of belonging. The local notables in Kilis embody the town's historic identity in their recollections of family biographies better than any other strata.

The traditional notables in Kilis constitute the old wealth that dominates the economic and cultural life as well as local politics till the 1960s. But the old wealth does not form a homogeneous group. It was crosscut by inner tensions, particularly between traditional landed notables and trade notables. The demarcation of the Turkish–Syrian border in 1921, despite succeeding amendments of the frontier line, meant that Kilis town had to abandon fertile agricultural lands and move to landed estates across the border at the Syrian side. It also shifted regional economy between Kilis and Aleppo, as well as cultural bonds of attachment. I focus in this chapter on the inner tensions and distinctions among local notables under the new historical context imposed by the border. Firstly, I introduce the notable families and describe their distinguishing characteristics with reference to the scholarly literature to provide a definition. This introduction clarifies the division among the notables and demonstrates that the traditional landed elite denied the status of notability to the trade notables.

The following sections point to the ways in which traditional landed notables act upon their vested interests in land and its political as well as economic consequences. Their desire to secure their landed estates encourages them to pursue their interests within the context of border disputes and negotiate the annexation of Kilis town with the French Mandate Syria. The consequence is the stigmatization of the notables as traitors and putting their entitlement to citizenship at risk. The disintegration of large-scale landholding and the loss of their landed estates at the Syrian side are coercive conditions that force the landed families to convert their capital held in one form to another in order to reproduce their social standing. The traditional landed notables consider it as decline.

Who represent the notables? Problems in identifying notable families

The access to local notable families had not been an easy process. As for my initial months in the field, I was challenged with several setbacks in identifying which families are viewed as the notables. It seemed unusual, because I assumed that these families constituted the local rulers in the Ottoman social order and they kept their power with the transition to Republican regime. Here, the ethnographic account of Michael Meeker (2002) about notable families of the town, a provincial district of Trabzon, is worth remembering. Meeker talks at the very beginning of his field research about how the conversations with local people and state officials led him to reveal that two families dominated the public life of the town over a century and this was a fact acknowledged to a certain extent by the locals as well (Meeker, 2002, p. 5). This is not the case with Kilis town. I will first introduce in this section a theoretical framework to discuss in what ways the notables constitute the upper stratum with reference to the scholarly literature on Middle Eastern historiography and social stratification.

One major setback in identifying the local notables is the changing demographics of the town. The urban center received a large number of rural migrants from nearby villages in the last four decades, while the notable families tended to move out of the town in order to seize better education and investment opportunities in the cities. The rural migrants do not distinguish the old wealth from the new rich in ascendancy in the 1970s and the following decades. The majority of town dwellers identify the wealthy families possessing economic and political influence as the notables. The young generations are brought up with stories about the notable families that they heard from their elders and embrace their elders' feeling of repugnance from the notables as their own experiment. The elders' recollections reveal a long-established tension between the rural and the urban until recently, which is rooted in the late Ottoman social stratification structure. The interviewees' recollections at the border villages reveal that they used to live in dire straits of sharecropping by cultivating the lands of notable families and the latter were influential patrons protecting the peasants' interests and needs. Thus, they help us to distinguish the old wealth and the perpetuation of their traditional paternalistic domination over the rural during the formative years of the Republic.

On the other hand, the urban natives of the town tend to contain the definition of notables as limited with traditional landed notables. The wealthy families of mercantile origin distinguish themselves from landed notables, calling them *eşraf* and name themselves as *esnaf* (tradesmen), even though they have been incorporated among the ranks of notables at the turn of the twentieth century.[1] Occasionally, families whose elders were assigned to higher posts within local bureaucracy might tend to conceal it by claiming their origin as farmer or tradesman, if their elders have made their fortune by abusing their authority or by getting involved in lucrative business like gold smuggling. In short, while the traditional landed notables and trade notables compose the old wealth of the town, the inner distinctions among them are maintained singling out the *esnaf* families (trade notables) from the *eşraf*. The following sections will elaborate further on the *eşraf* identity as well as the political contention at the historical background of the distinction between *eşraf* and *esnaf*. This distinction is important to understand why traditional landed notables, rather than trade notables, experience a strong sense of falling from grace.

Question of notables as early capitalists in Kilis

This chapter draws on the "politics of notables" as a paradigm to explore the old wealth and their social mobility strategies in Kilis town. The notion "politics of notables," asserted by scholars of Middle Eastern history such as Hourani, Khoury, and Dawn, relies on the fact that the urban politics of the Ottoman provinces can be understood if seen in terms of a politics of notables or to use Max Weber's phrase, a 'patriciate' (Hourani, 1993, p. 87). The local notables derived their sources from their control of land and land tax, urban real estate, local handicrafts, trade, and *evkaf* (*waqf* properties). The notables implied a political class in the Ottoman Empire, as they could play a certain political role as intermediaries between government and people and—according to Hourani, within certain limits—as leaders of the urban population.

In the Ottoman historiography, the notables can also be called *ayan*, referring to the local governors assigned by the central authority with the duty of tax farming for the Ottoman treasury. The notion of *ayan* began to be used for the local governors identified with the abuse of the ruling authority, public office, or tribal power for arbitrary government, imposition of extra taxing, and embezzling waqf property with the deterioration of tax system and mostly signified a certain era in the Ottoman history. The alternative notion of *eşraf*, derived from the Arabic plural of the word "*şerafet*" (the dignified ones), does not have such a political connotation or historical reference similar to the notion of ayan, but it is equivalently used for signifying the notables. However, both notions are value-laden concepts and ascribe a moral quality or essence.

To commence my discussion about the notables as early capitalists, I should first introduce the question how the Hourani–Khoury paradigm defines the notables as a class in sociological terms. The significance of this question is twofold: first, it will provide me with a theoretical framework to understand the social stratification in Kilis town inherited from the late Ottoman Empire and the continuum between two different phases of history, namely the Ottoman and the Republican era and second, it will help me to explore how the local notables in Kilis reproduced their social position.

When Khoury (1990) revisited his own discussion about the paradigm, he reexamined it in terms of a class analysis based on his research on Syrian geography. For

Khoury, the class analysis could be helpful for understanding the nature and behaviors of urban notables after the introduction of the 1858 Land Code which allowed private ownership of the land and the rise of a landowning-bureaucratic class. Here, the class was defined in terms of its relationship to the means of production as well as to its social position of its constituents (peasants, artisans, etc.). However, he asserted that despite their ascendancy, the notables could not be sharply defined as a class formation in the late Ottoman era since the private ownership of the land was not yet consolidated until the French Mandate and the evidence available did not "justify the conclusion that property rights became the main source of power" (Roded, 1986, pp. 380–381; quoted in Khoury, 1990, p. 221). In other words, the notables could not be defined as a class in a full Marxian sense.

Therefore, two points are important to define the notables as a class. First, the notables evolved from the late Ottoman traditional order to a modern one. For both Khoury (1990) and Mardin (1967), the private landownership plays a significant role in changing the social stratification structure of the Ottoman society and characterizes this transition. Whether the private landownership on land fostered commercialized large-scale agricultural production is another question that I will address later on. Second, the private ownership of land is not sufficient to define the notables. The notables also benefited in holding offices in local bureaucracy and engaging in patron–client relationships in order to sustain their social standing or status in a hierarchy of power and domination.

Although these approaches indicate that the notables are an "Ottoman" element, they suggest the continuity of notables' significance after the demise of the Ottoman Empire. The notion of "politics of notables" is particularly used to denote the continuing significance of the notables as a class in the late Ottoman and post-Ottoman hierarchy of power and domination, and allows the investigation of the incorporation of new actors in its perpetuation.[2] Not all scholars agree with Khoury and Mardin about the perpetuation of the notables in modern Turkey. The historian Keith Watenpaugh argues that politics of notables perpetuated in French Mandate Syria but it disappeared as a viable technique to comprehend center–periphery relations in Turkish Anatolia as of the mid-1920s—except perhaps in parts of Kurdistan (Watenpaugh, 2003, p. 278). On the other hand, studies of local politics in the formative years of the Turkish Republic demonstrate the penetration of traditional patron–client relationships to the party politics under the single party regime (Sayarı, 2011). These relationships took the form of political clientelism and party patronage that played a greater role after the emergence of multi-party politics.[3] Although the extent to which local notables maintained their domination within the context of party patronage should be explored further, these studies allow extending the politics of notables as a viable paradigm to the Republican period. This chapter does not aim at analyzing center–periphery relations through the politics of notables but it intends to show that the notables remained influential in Kilis border town until the 1960s. Hence I have enlarged the scope of Watenpaugh's statement to Kilis town and argued in this chapter that the politics of notables in Kilis town is perpetuated by the incorporation of a new middle class in the 1920s and the traditional landed notables maintain their position in the social hierarchy though they had to share their power and domination with this emergent middle class.

Now, did the private ownership of land pave the way to the large-scale commercialized agriculture and to the emergence of notables as early capitalists? This is a highly contentious debate in the scholarly literature. The main contention area is the nature of land tenure in the late Ottoman era. Extensive historical research indicates that the late Ottoman agricultural system was dominated by small-scale subsistence farming and the large-scale landholding was exceptional because possession of state-owned (*miri*) land was not actually allowed (Keyder, 1991). An exception to the Ottoman land tenure was the evolution of *çiftlik*, farms that turned state-owned land into privately-owned properties in the nineteenth century. They evolved when the Ottoman treasury started leasing out state-owned land by public auction in order to overcome its financial constraints. But this was the case of opening up uncultivated and waste land, where *çiftlik* remained again the legal property of the state.

Among the landlord-managed estates, the large-scale commercial exploitation was seen only in certain regions which are particularly exposed to the enlarging world economy. Halil İnalcık suggests that the plantation-like farms relying on the exploitation of peasants as sharecroppers evolved where "leaseholders were economically motivated to maximize their revenues under the impact of an expanding external market" (İnalcık, 1991, p. 113). The tax and rent collecting landlords were able to "enclose" after the land code of 1858 in exceptional cases like Syria and Iraqi provinces. Thus, the conditions of agricultural production and landlord–peasant relationship were not generally transformed into an exploitative one. Drawing on İnalcık's study, Çağlar Keyder argues that the independent status of peasantry was largely protected. Agricultural exports, unexpectedly, did not originate from the landlord-managed estates, but were derived from peasant surpluses.

The opponents of this view assert landholding notables as an exploitative class as 5% of the families living in the countryside owned 65% of the total land at the beginning of the new Republic (Köymen, 2009, p. 26). Commercial agriculture was a particular source of wealth during the wartime period between 1938 and 1945 and it had been the large-scale farmers who benefited most from the agricultural support programs rolled out in the following decades. We are reminded by the fact that the landholding class continued to seize peasants' surplus by using non-market mechanisms as well. The landholding notables were in relationship with the peasants as merchants and moneylenders and the peasants could never manage to subsist independently. Most peasants were tenants or sharecroppers on landlord-managed estates and half of their crop was seized by the landlords using their paternalistic domination and economic and political influence. Moreover, the amount of land owned by most of the peasantry was so small that it did not suffice for small-scale subsistence agriculture and made the peasants dependent on the landlords.

So it is plausible to question whether agricultural production was commercialized in Kilis town. Large-scale landholding was no exception in this region. Private possession became prevalent especially after the 1858 Land Code that recognized the existing *de facto* distribution of land ownership (Karadağ, 2005, p. 62). As mentioned before, oral and written accounts of Kilis locals indicate that the notable families seized private possession of large lands by bidding public auctions of tax farming and 'usurping' the land of debt-ridden cultivators. Research shows that investing in land became a rampant tendency among merchants and money-lenders who wished to enjoy the status of landowners in northern Syria (Bouchair, 1986, p. 102). Thus, 70% to 80% of the villages in the region belonged to large-scale landowners by the early twentieth century. Anecdotes of elder peasants that I listened to at certain border villages support the existence of large landholding. They highlight that peasants particularly in fertile lowlands used to labor on the landlord-managed estates rather than rely on subsistence farming. As sharecroppers, they had lived in dire straits till the 1960s.

However, there is no evidence to show that organization of agricultural production was totally altered. The city of Aleppo and its rural hinterland was exposed to external markets. The city of Aleppo became fully enmeshed with world capitalism towards the end of the nineteenth century, which made the city further integrate into the economy of the Ottoman heartland in Anatolia (Masters, 2010, p. 292). But, international trade was largely held by non-Muslim merchants. Agricultural trade between Aleppo and its hinterland remained internal. Commercialization of agriculture made land a lucrative source during the nineteenth century and not only local notables, merchants, and highly placed officials, but also upper class peasants acquired land (Karadağ, 2005, p. 62). The notables of Aintab (today, Gaziantep), for instance, did not adopt a market-oriented production in their estates, neither did they invest in modern techniques of farming (Karadağ, 2005, p. 110). Notable families in Aintab retained most of their revenues from landed *rentiers*. Thus, the private ownership of land seems not to initiate a shift in the organization of agricultural production.

Still, the distinctive feature of local notables in Kilis is their ability to diversify their revenue-bringing ventures and their access to the trade markets in the post-Ottoman Greater Syria where they continued to sell their crop. As early as the 1940s, the interviewees' recollections reveal the illegal trade of gold as well as other products across the newly demarcated borders with Syria. The official documents and minutes of the national assembly document the efforts of state authorities to press the issue as "smuggling" and take preventive measures during the early Republican period. These archival sources as well as interviews demonstrate that members of local notables could circumvent the border regulations by abusing their rights of crossing the border and gathering their crop afforded to them as entitled landholders. On the other hand, the border peasants were seldom immune to the charges of smuggling.

In conclusion, the local notables constituted the old wealth until the rise of the new rich in the 1960s. They did not fit into the definition of the capitalist class in a full Marxian sense because large landholding did not alter the production mode. Nevertheless, as I have discussed in the following sections, landed estates in Syria offered landowning families the chance to seize the opportunities of investing in trade, particularly through 'illegal' means. First, they could sell their produce in

Syria for higher prices by circumventing the border transit and customs regulations. Normally, the border regime allowed them to harvest their produce at the Syrian side on the condition that they would bring them back along the border of Kilis and sell them on the domestic market. Second, they could engage in the lucrative gold trade, by bringing the gold clandestinely to Kilis. In this vein, I have argued below that local notables were early capitalists insofar as they could use the border transit regime as a source of economic accumulation and seize new opportunities in the circulation of products and money.

Eşraf *identity in Kilis town: traditional landed notables*

The *eşraf* identity among the urban middle strata in Kilis town implicates the traditional landed notables with vested interests in the Ottoman social order. It does not include the trade notables and the new wealth that emerged in the 1960s. This section introduces *eşraf* identity from the viewpoint of town dwellers. The dwellers held morally laden and seemingly contradictory definitions of *eşraf*, regarding them as both despots and modernizers. I assume that their accounts help to a better understanding of why *eşraf* identity remains limited with the traditional landed notables.

The interviewees among the urban middle strata agree that traditional landed notables composed "the *eşraf* system" in which certain families were interconnected mostly with kinship and marriage relations and were headed by feudal landlords who maintained a prosperous urban life due to their domination over the agrarian countryside. According to them, the *eşraf* system came to an end in the 1960s. That is to say, the *eşraf* status lost its distinctive attributes that used to characterize the traditional landed notables. Also my interviewees among traditional landed families admit that the distinction between themselves and tradesman families was abolished in the 1960s. These aspects will be further explored in the following sections in relation to the disintegration of large-scale landholding. It suffices to say that with the *eşraf* families, it is understood the traditional landed notables composed of highly placed military-bureaucrats, ulama, and secular leaders "whose power might be rooted in some political or military tradition, the memory of some ancestor or predecessor; or in the *asabiyya* of a family or of some other group which could serve as its equivalent; or in the control of agricultural production through possession of malikanes or supervision of waqfs" (Hourani, 1993, p. 89) in the late Ottoman era. This definition of *eşraf* assumes the persistence of this status until the 1960s and excludes the wealthy families of mercantile origin as well as other well-educated professionals, artisans, and bureaucrats that emerged on the eve of the new Republican regime among the ranks of notables.

The most elaborate opinions about *eşraf* are expressed in the works of local intellectuals from the urban educated middle and upper strata, whose generations witnessed the late Ottoman or early Republican period. This is particularly true for the *History of Kilis* by Kadri Timurtaş, published in 1932, which includes eyewitness accounts of their social milieu. Timurtaş, a lawyer and chronologically-distant descendant of a high-ranking governor, served in the Ottoman court and addressed in his work the social and political history of notables. Yet, his work is also helpful in revealing the subjective opinions and moral judgments about these strata as well as providing historical background on the basis of factual information, even if not within a scholarly discipline. According to Timurtaş, large-scale landholding is the ultimate distinguishing characteristic of *eşraf*, but it is not the only one. As mentioned in the previous section, families of tradesman origin acquired large lands by buying the low-price bids after the Land Code of 1858. The *eşraf* are also regarded as landholding despots (*mütegallibe*), leaning towards seeking dominance, usurpation or unfair profiting, and patronage.

His definition of *eşraf* replaces the class of ayan that used to refer to the military-bureaucrats assigned by the central authority as district governors—such as *sancak beyi, voyvoda,* or *mütesellim*—and heads of cavalry (*mutarasarrıf*) entitled to appropriate the land and liable to collect tax.[4] The *eşraf* families did not hesitate to take away the land and real estate forcibly from the people. It was likely that they would seize a village in which they stepped out by purchasing a small plot of land. Their sole occupation was leading the way of their adherents to the governmental offices and getting their affairs done. They used to buy up the tax auctions cheaply and thus obtain large amounts of land. His point of view is far from neutral. *Eşraf* members, for Timurtaş, even seemingly intimate and polite in their relationship to each other, were in effect hypocrites because they detested each other in pursuit of their personal interest and in competition among themselves.

On the other hand, *eşraf* implied a moral quality or essence for Timurtaş as well. The above-mentioned characteristics pointed to what *eşraf* had become rather than what it really was. As he writes, the *eşraf* members in Kilis town were people embodying an inborn nobility, and intellectual and moral cultivation (Timurtaş, 1932, p. 86). They had a refined tradition of personal dignity and kindness. Generosity, caring for the poor, and complaisance with the governmental authority constituted major signs of notability (*şerafet*), derived from the Arabic word *eşraf*. But eventually, as Timurtaş assumes, the notion of notability lost its original purity.

The ambivalent opinion about *eşraf* put forth by Timurtaş is still prevalent among the educated urban middle class in Kilis, though the latter does not ascribe a moral essence to the *eşraf* identity. While my interviewees refer to *eşraf* as despots, they also indicate the dignified and reputed people who assumed a progressive role among the town community. They refer to the erudite and learned members among the *eşraf* in order to sustain their viewpoint and mention about their role as educators in the town. Most scholars and teachers in late Ottoman religious establishments were recruited among the *eşraf*. The known turcologist Necip Asım Yazıksız writes about her childhood memories[5] that the town was a seat of poets and scholars teaching in local madrasahs where he took lessons. The town accommodated several schools[6] providing religious and scholarly education at the end of the nineteenth century and the logic courses in the town earned such a reputation that they were attended by students from elsewhere. As Timurtaş's account reveals, the nineteenth century saw the rise of ulama families in the town who gained reputation and recognition as notables due to their advancement in scholarship and wisdom (Timurtaş, 1932, p. 90).

The urban middle class in the town believes that *eşraf* led the town community in embracing the Kemalist modernizing reforms and eased the transition from the old dynastic order to a new republican regime. Because *eşraf* members were well-educated, they could adopt to the reforms more readily (Çolakoğlu, 1991). I suggest that *eşraf*'s acceptance of the reforms stems from a high degree of adaptation to changes in its relation to the central authority. The scholarly literature indicates that the transformation from the Ottoman state into the new republic on the periphery was not without contention. The authoritarian adherents of radical reformism and Westernization achieved dominance over others in the bureaucracy and imposed the legitimacy of their rule on a principle other than religion and dynasty by oppressing the opposition (Keyder, 1999; cited in Durakbaşa et al., 2008, p. 24). Though the political contention between the *eşraf* and the emergent notables in Kilis inflicted a heavy blow on the *eşraf*, it continued to maintain their social position among the town community. The ulama families and descendants of the dynasty maintained their prestigious place among the *eşraf*, particularly with reference to their pioneering role in adopting modernizing reforms.

Ottoman religious schools of the town are regarded as the trademark of the town's modernity during that era, although many of them are not preserved in terms of architectural heritage. This is particularly significant, reminding that the Republican reforms have discredited these establishments with a closing act in 1925. Particularly, the old naqshbandi lodge attests the modernity of the town in the eyes of local dwellers. Built in 1858 and losing its function in 1925, the courtyard of the lodge where the tombs of two successive sheikhs are located is flocked every Fridays by local visitors, mostly women who come to pray for their wishes and dole out food to the poor. The late sheikh, educated in medicine, was known to be adept in poetry—his wife was reportedly adept as well. But his son took over after his father's death and he is most popularly known. He is remembered as a pioneer in welcoming the hat reform outlawing religious headgear by putting on a hat right away. Thus, the lodge would not be closed by Atatürk, the founder of the state. An interview with a member of the traditional landed notables[7] illustrates how the sheikh is regarded not only as a pioneering figure but also as an influential one. The interviewee mentions that the sheikh is his grandfather, since the sheikh and his family are related through bonds of marriage:

> For example, when the hat reform in Turkey was first introduced in Kastamonu city, all mayors and muslim scholars of Kilis gathered upon the news that someone called Atatürk banned the headwear and replaced it with hat. My deceased grandfather was already prepared. He had very good relations with Atatürk after all. One of the two lodges that were not closed in Atatürk's time in Turkey is this one, the other one being the lodge of İsmail Agha community in Istanbul. My late grandfather, reacts that what you call hat it is just this at the end and taking off his headwear, he puts on a felt hat. After that, the hat is used first and the utmost in Kilis town compared at the country level since

the sheikh effendi had put it on. My grandfather, the late, put on that hat till he died.[8]

Actually, the sheikh, taking over from his father, ended his religious mission himself and stopped to hand it down to his successor in 1925 so as not to contravene the closing act (Şahin, 1999, p. 27). But my respondents attribute such a great reputation to the lodge that they wrongly remembered it as remaining open after the closing act of 1925.

The transition from the Ottoman social order to the new republican regime brought significant changes in the social composition of notables. The emergence and incorporation of a middle class of professionals, bureaucrats, and traders into the politics of notables was made possible under the conditions fostered nationwide by Kemalist cadres. The new regime sought to create a national bourgeoisie by eliminating non-Muslim elements of international trade. National modernization is led by the pursuit of radical reformism and Westernization, while purging religion and the dynasty elements of the old order. The following section explores in what ways this political transition formed a background to the distinction between *eşraf* and *esnaf*.

Historical background to the distinction between eşraf *and* esnaf

The period between World War I and World War II witnessed the rise of a new middle class of educated professionals, bureaucrats, merchants, and moneylenders in Aleppo and its provinces. Concerning Kilis town, this new middle class emerged and it was incorporated into the politics of notables thanks to the Independence War fought against the Entente States as well as the Ottoman Sultanate. While this time period gave way to the emergence of political patronage relations under a single party-led government, the *eşraf* tried to contain and control the rise of the emergent middle class, which in turn distinguishes itself from the traditional landed notables by embracing their mercantile origin. This section explains that the distinction between *eşraf* and *esnaf* or, preferably, traditional landed notables and trade notables, is rooted in the political contention among themselves that emerged during the Independence War and its aftermath.

The political contention typically took place within the context of the transition from the Ottoman social order to the new Kemalist regime that is supported by the emergence of a new middle class. Historical studies on the Mandate-era Syria shows that an urban middle class composed of lawyers, doctors, schoolteachers, bureaucrats, international merchants, bankers, and their families rose among the notables in fighting against the French Mandate (Watenpaugh, 2003, p. 259). The wave of Arab nationalism rose in the interwar period among the intellectual circles of Aleppo and backed this new middle class in their incorporation into the politics of notables. But they were nevertheless contained and controlled by the traditional landed elite. This is also relevant for the rural hinterland of Aleppo.

In Kilis town, the period between World War I and World War II witnessed the rise of a new middle class, which constituted the trade notables. This period also reflects the development of political patronage relations as the single-party regime led by the Republican People's Party (RPP) sought the support of notables (Sayarı, 2011). The emergent trade notables were also mobilized into politics of notables, though the traditional landed notables succeeded at large to contain their vertical mobilization with the backing of local bureaucracy. Since it is the accounts of the Independence War that makes political contention between the traditional landed notables and the trade notables visible, my discussion will focus on the wartime history.

Research on local notables conducted in four Anatolian cities observes that the Independence War is regenerated in the formation of urban identity, spatial organization, and urban iconography, while it becomes an important reference in the discovery of local history (Durakbaşa et al., 2008, p. 26). The visit of Atatürk to the town while he was the 7th Army Commander in World War I[9] is remembered and mentioned in my conversations with the notables and educated urban middle strata as a significant event that started local resistance. His words about the astuteness of Kilis locals at this visit is popularized as a source of pride and frequently referred to in these conversations. The narratives told about the liberation from foreign occupation are incorporated into the formation of the town's identity.

A wartime history written mainly by local historians/scholars[10] is popularized through extensive coverage in local newspapers and websites such as the local publicity pages of governorate, and occasional academic events. These written accounts of wartime actors and state archives—usually of General Staff—are nationalistic accounts that start Republican history with the Independence War. However, they neglect the ideological and political foundations of the new regime in late Ottoman history. These accounts do not give clues of how the Committee of Union and Progress (CUP) led a provincial 'grassroots' politics, particularly organizing the artisans and traders around Turkish nationalist tenets (Canefe, 2002, p. 144). Thus, when the Unionist regime needed to eliminate non-Muslim populations which held trade relations with international markets in order to ensure the creation of a national bourgeoisie (Keyder, 1987), the emergent trade notables as well as the traditional landed notables would support the Unionists in the province. There are reasons for assuming that these strata are involved in extending the Unionist policies of Armenian deportations to Kilis town and seizing their abandoned property.[11] In short, by the time of the Independence War, the traditional landed notables consisted of Muslims, together with a few wealthy merchants among the Jewish population in the town.

The town saw successive occupations by British (1919) and French (1919–1921) troops at the end of World War I. The war accounts in written sources depict factionalism between the traditional landed and trade notables, beginning with the mobilization of local resistance against the French occupation. Initially, both traditional landed notables and trade notables sponsored the local militia against the French. The Islamic Union (*Cemiyeti İslamiye*), for example, was organized in several cities and districts of Ottoman territory.[12] However, the French occupation following the short acquisition of the town by British troops marked a divergence among the notables. The armed resistance recruited from aghas and merchants, as well as several members from the traditional landed notables. Yet hesitations or distrust dominated among the latter and they abstained from joining the armed mobilization. Families moved to Aleppo during the heyday of occupation in order to be secure. Thus, the traditional landed notables are considered as collaborators of the Entente states or deserters in war.

For Latife, an old-generation woman from the traditional landed notables born in 1911, wartime and its aftermath represent a period of rift among the town community. She calls the period "the time of Kuvayi Milliye" and she believes that the hatred among the common people against their seniors did not exist before. The time of Kuvayi Milliye serves as a litmus test for distinguishing the traditional landed notables from the trade notables who gained power and influence with their wartime mobilization.[13] The latter's children today believe that their parents represented the "people" as opposed to the deserter traditional landed notables. The divergence among the traditional landed notables and emergent notables evolved into political contention in the aftermath of French withdrawal: the power struggle over controlling the candidates for parliamentary politics, executives in local bureaucracy and party politics as well as local associations.

The political tension escalated when a school principal and a doctor who did not join the armed mobilization in Kilis were found guilty of the charges and were dismissed from public service. The Law Article 854 about the civil servants who did not join the national struggle was proposed by Ahmet Remzi Güres, a war veteran from Kilis elected as Gaziantep deputy and his colleague Mazhar Müfit Kansu, another war veteran and deputy (Çolakoğlu, 1991). Though the charge against the doctor was eventually dropped by a higher deciding committee reviewing the investigation in 1928, the conviction of the school principal was conclusive. The school principal was influential in curbing the power of his opponents in taking over the board of local associations[14] and local governing bodies such as the mayoralty and provincial council. He was influential in bringing the traditional landed notable members in power within the local party branch of RPP as well. His struggle to restrain the ascendancy of a wartime hero, a member of the new notables, would be effective, as the city and district governors would back him up. He would also be supported when he forced his opponent to step down—though elected—from his office as mayor and be replaced with a new election by a traditional landed notables' member. Thus, he became the target in the midst of the political contention incited between the traditional landed notables and new notables. The wartime history and its stigma on the traditional landed notables conceal the emergence of a new middle class into the ranks of notables in the popular imagery of the town community. These new notables were particularly empowered by their mobilization in the local militia and successfully draw political support upon patronage relationships established under the rule of the single-party regime. Still, both factions of notables continued to come up with mayors and parliamentary deputies in the following decades. These relationships would help them to maintain their ties with the central authority of the newly-founded Republic and further strengthen their social and economic influence in the province. While the new notables took support from the rising wave of Turkish nationalism, they presented themselves as representatives of the people opposing the rule of old notables. Therefore, the popular imagination of *eşraf* remained intact and limited with the traditional landed notables despite a new social composition.

The story of an *eşraf* mansion: unwanted citizens, muted histories

There is a strong consensus among the town community of silencing the unsettling aspects of wartime history. Long after starting my fieldwork, I realized that despite the prominence of memories about the Independence War in the references to local history and the urban identity of the town, my interviewees selectively detached certain controversial events and never mentioned them. I traced unspoken aspects of wartime history upon the controversial story of a mansion transferred from a traditional landed family to another family of trade notables. The memories that condemn members of traditional landed notables of treason during the French occupation at the end of World War I (1919–1921) have been muted in order not to offend the notables' descendants. The traditional landed notables tried to claim their vested interests in their landed estates left across the newly demarcated border in the midst of warfare and they eventually faced the risk of denaturalization and confiscation of their properties. Although very few of them had actually been convicted by the new Kemalist Regime, the traditional landed notables are still remembered with this stigma of being a traitor. The silencing of unpleasant memories further stigmatizes their parents as collaborators of enemies or traitors, rather than making them fall into oblivion. My discussion asserts that wartime memories constitute a material force in the present to question the traditional landed notables' claims of hegemony, while also putting out of sight the property transfer between traditional landed notables and trade notables.

The "National Struggle" fought against the foreign occupation of the town is central to the conversations with the dwellers about the local history. Border studies show by drawing on the notion of collective memory: that personal memories are always constructed and located in the social context. Border regions are particularly crucial sites for the recovery of national memory, as well as its contestation and re-negotiation (Zhurzhenko, 2011). Yet, the retrieval of memory is selective and the construction of collective memory implies a consensus on what should be remembered and how. The wartime history revolves around the heroic resistance fought against the French troops. These memories draw on a 'patriotic Turkish nationalism' where the Independence War remains pivotal to the founding myths of the nation.[15] Thus, it discriminates between the participants of local armed resistance and the deserters, by putting its stigma on the latter.

The disclosure of this stigma was possible as I revealed the story of a mansion that had been first confiscated from a traditional landed family and then bought by a member of the trade notables. I visited this mansion in order to interview the heir to the trade notable family. It was a two-floor stone house with a courtyard and elegant details, built in the traditional architectural style of the region. My interviewee Şükrü was polite in showing me around as the house was recognized among the fine architectural examples of the town's cultural heritage. Şükrü's family had owned it since 1926, when his grandfather bought it.[16] Long after I interviewed him, I discovered that the house had a different story that was not revealed to outsiders. According to hearsay, the house was confiscated from the old proprietor family when a member had been charged with treason by the notorious trials of the Independence Tribunal (*İstiklal Mahkemesi*) in the aftermath of the Independence War. The old proprietor was performing his duty as a district governor under French occupation. He was included among the list of 150 convicts who had been subject to denaturalization and their properties were expropriated. He had to desert the town and live in Aleppo.

The controversial story of the mansion was intriguing enough to attract my attention. Curiously enough, none of my interviewees mentioned it. As referred to later, I talked to several members of this family. Only Nihal, the bride of the family in her mid-sixties, had mentioned about the mansion. As a daughter of a governor at a Western Anatolian city who came to Kilis only after her marriage in 1975, she was rather assimilated into town life and she continued to live even after her husband's early death. She was sensitive about the recognition of the house by her husband's family name and claimed the property. She even showed me the title deed but did not ever mention why the house passed into other hands. Ironically, her father-in-law's house, also an old stone mansion where she continues to reside, stood back to back to their old confiscated house.

The family, from traditional landed notables, had several mansions with indoor and outdoor passages connecting them and they almost formed a bloc of buildings on a parcel surrounded by streets. Being neighbor to the new proprietors appears to have kept the feeling of offense in having lost their property for unjustified reasons. At least, this was my impression. For instance, I observed that she kept an eye, intentionally or unintentionally, on the modifications and repair work on her family's old house made by the new proprietors as she had mentioned it during my visit to her house. As I visited Şükrü, the current proprietor of the house, he hinted at how the roof insulation work had become a concern for the neighbors and the municipality officers came to check the removal of old Marseille tiles.

Another member of the old proprietor family said that the house was spoken of as an abandoned property of Armenians—yet another intriguing aspect of the story that I would not be able to explore further. But there had been no conversation about the family's dispossession of the house. Not only the family, but also other town dwellers abstained from speaking about it. But these memories still impinge on daily life powerfully. An interview with Celal, a self-taught historian born in 1963, proves that the town dwellers tend to regard the disclosure of these memories almost as scandalous. I visited him at the university library where he was employed and, thanks to his job, he had access to the Ottoman manuscripts, which were mostly religious texts with a few exceptions of diaries. The university library aimed at compiling a documentation center on local history and received book and manuscript donations from the locals. As I asked him about local notables, he recalled a testimony about the notables' cooperation with the Entente Powers written in a diary manuscript. The diary recorded a meeting during the British Occupation of Kilis[17] held by the American Red Cross Commission[18] with the local notables in the town. He would not be able to show me the manuscript, since the center would not reveal these resources in order not to offend the grandchildren of notable families. Still, he mentioned about the author's feelings of resentment when he witnessed the notables' wishes from the committee to make sure that Kilis should stay annexed to Aleppo for the security of their landed property.

Though the strong ties with Aleppo were not severed, the town was annexed with Aintab (Gaziantep) district in 1911. But the controversies over the administrative division were principally caused by the new geopolitical situation in the Middle East. The Entente Powers had foreseen the establishment of an Arab state in Ottoman Syria and Iraq under their auspices. Thus, the desire of notables to secure their landed estates in a country under the auspices of Entente states would be regarded as cooperation with external powers and a disgraceful intention. My encounters with members of notable families provide insight into what extent they could be affected by the stigma of their elders, as the wartime allegations of collaboration grew serious.

The old proprietors of the mansion had another relative targeted by the similar charges: the school principal mentioned earlier. He held his post during French occupation and was found guilty of desertion from the Independence War. When I met his grandchild Faik, a member of traditional landed family in his early sixties, to interview I did not have any knowledge about it. He was very cautious in choosing his words and, yet, he did not let me record or write down our conversation. As I explained him about my intention to learn about family history, he offered to provide me with a "general framework." Then, he mentioned about his grandfather's memoirs from his unpublished diary. He said that his grandfather was charged since he was carrying his administrative duty during the occupation. Yet, his grandfather was among the few influential figures of local resistance and another member from his family, a timar holder who commanded 500 cavalry against the French and had great contributions. My interviewee would recommend to me that I should take into account these contributions as well in order to make a thorough evaluation of wartime history. I should also add that he would not be able to meet my request to have a copy of the diary.[19] The way in which he received my demand for an interview made it clear to me that this notorious past could be a moral burden for the family.

My interviewee's grandfather was in fact a member of Mar'ashli, one of the large landholding families of Aleppo that took part in the Syrian National Congress at the end of World War I, fighting for independence against the French Mandate (Watenpaugh, 2005). He had a rural mansion in Kilis, where the family held large lands. The anecdote about his heroic participation in local resistance against the French probably referred to his activities as part of the secret committee of the Islamist Union, forming nationalist gangs to fight the French army in the early 1920s (Mizrahi, 2003). These gangs made raids on French Army bases in north Syria and took refuge at the Turkish side of the border. The stigmatization of notables with nationalistic accounts makes these histories muted and unintelligible. These nationalistic accounts also disregard the fact that the family histories of local notables are straddling the border and it is difficult to categorize them as part of 'national history.'

Apparently, several other members of traditional landed notables faced allegations of treason for deserting local armed resistance against the French troops during the occupation as well. These people had to leave the town because of their fear of conviction and they could be at ease after the 1938 Amnesty. The town elite tended not to speak about wartime memories which could be embarrassing for the traditional landed notables. But an interviewee of mine, Ekrem, a local historian born in 1935, brought up the issue. He reported about a member of the traditional landed notables who had to desert the town and then came back. He mentioned that he was a traitor and named him within the list of 150 convicts—in fact, only the former district governor of the town was involved in the list:

> Now, after Atatürk fought the Independence War, they made up a list in Ankara in Republican era. There were 150 people. There were people not only from Kilis, but also from Istanbul, Ankara and everywhere in Anatolia where betrayal of the country occurred. I even saw someone. He came after the war [possibly World War II]. He used to publish a local newspaper here, Hududeli Newspaper. I had talked to him and he said he did not betray. Of course, there is contention back then, so supposedly the commanders coming here expelled him in order to seize his properties.[20]

As the above quotation illustrates, despite a small number of convictions, the image of the traditional landed notables as collaborators of the enemy remains. It seems that wartime mobilization questioned the claims of traditional landed notables to a position of hegemony in the town with reference to their desertion from the Independence War. Even though the traditional landed notables continued to enjoy their status of notability together with the trade notables, the wartime allegations of collaboration left its mark on them. Two members of traditional landed notables are known to have survived convictions and legal penalties such as denaturalization, confiscation of property, and a ban from official duty. Still, this stigma is generalized to the traditional landed notables who had deserted the armed resistance against French occupation. The wartime memories are likely to contribute to the traditional landed notables' experience of falling from grace. They also hide from sight the transfer of estates between traditional landed notables and trade notables. Whether there was any abandoned property of the deported Armenians among the transferred estates is another question that could not be addressed within the scope of this chapter.

Concluding remarks

This chapter has focused on the historical context imposed by the border and in what ways it shifted the class and status relationships among local notables. As traditional landed notables tried to influence the demarcation of border upon their vested interest in land—their properties were mostly located in north Syria—and demanded to be annexed with Aleppo at the fall of the Ottoman Empire, they became accused as traitors and engendered by the risk of losing basic citizenship rights. This period had a deep impact on traditional landed families, causing several members, accused or under investigation, to go to Aleppo and divide among close kin. On the other hand, the Independence War and its local mobilization created the conditions for an emergent middle class to become involve in the politics of notables and helped families of mercantile origins to gain political power. These families, as trade notables, shared power with traditional landed notables, though they could not topple them in the local bureaucracy and party politics.

This chapter has also discussed whether local notables could be designated as early capitalists in relation to the nature of land tenure. Local notables could benefit from a border transit regime in order to reap the harvest on their landed estates at the Syrian side on the condition that they would bring their produce and sell in the domestic market. The discussion concluded that large-scale landholding did not result in the commercialization of agricultural production, but the notables capitalized on the border transit regime to yield differential revenues from the selling of their produce in foreign markets and engage in gold trade.

Notes

1 Beyhan, an old-generation woman from trade notables, has particularly emphasized the distinction between *eşraf* and *esnaf,* reclaiming her family's mercantile origins.
2 The notion is inserted by Hourani with his 1968 "Ottoman Reform and the Politics of Notables" article referred to earlier. The most consistent form of the politics of notables was found in Greater Syria and Hijaz (Gelvin, 2006).
3 As the landholding notables climbed up along the vertical ties of party politics, they were more strongly represented in the parliamentary system. The party-directed patronage allowed it to rule over the peasants by being influential over the distribution of public resources to them as "favors" like employment opportunities, favorable treatment from state officials finding medical care in Ankara or collective goods such as infrastructure building and price subsidies for agricultural products (Sayarı, 2011, p. 9).
4 See Kıvrım (2008) for the evolution of ayan in Kilis.
5 Yazıksız, born in 1861, says that he left the town for further education in Damascus and Istanbul in his thirteenth year. His childhood then refers to the years between 1861 and 1874. This short memoir was published in the *Türk Yurdu* periodical in 1927.
6 Kilis had 1,456 Muslim students in primary schools (*sıbyan mektepleri*), 75 students in the secondary school (*rüştiye*), and 108 students in 24 madrasahs in the years 1891/1892. There were also 285 students in seven non-muslim schools in the years 1891/1892. See Aleppo in *Ottoman Provincial Annuals* edited by Eroğlu, Babuçoğlu, and Köçer (2012).
7 Rauf, born in 1951, is a descendant of a former local governor.
8 *Mesela Türkiye'de, Kastamonu şehrinde şapka devrimi ilk olduğunda, bütün Kilis'in belediyeleri, hocaları toplanıyor: Efendi efendi, Atatürk diye bir şey çıkmış sarığı kaldırmış şapkayı yapmış diye. Dedem rahmetli hazırlıklı. Zaten Atatürk'le ilişkileri son derece iyi bir kişi. Türkiye'de Atatürk zamanında kapatılmayan iki tane tekkeden biri bu, bir tanesi de İstanbul'daki İsmail Ağa Cemaati. Dedem rahmetli "ya çok garip şapka dediğiniz de şu değil mi?" diyor. O zaman sarığı çıkarıp fötörü takıyor. Ondan sonra Türkiye'de ilk ve en çok şapka Şeyh efendi taktı şapkayı diye Kilis'te kullanılıyor. Dedem rahmetli de ölünceye kadar o şapkayı giyerdi.*

9 Mustafa Kemal (Atatürk) came to Katma station annexed with Kilis, as the Ottoman army lost the war against the British on the Syria battlefront and met the notables in Kilis to convince them to mobilize a local resistance in 1919. Historical resources record it as the beginning of the first militia groups (İnce, 2004).

10 Çolakoğlu (1991), Beşe (2009), Gülcü (2012), İnce (2004), Öztürk (2005).

11 See Kevorkian (2011, p. 610).

12 Gülcü (2012) denotes that the organization, established in Berlin, had branches in Anatolia, Syria, and Egypt. The local branch in Kilis first mobilized the militia groups watching for the security of neighbourhoods (İnce, 2004).

13 According to Çolakoğlu, the political contention among the notables was a conflict of interest that has been alienating for the lower strata and with long-lasting destructive effects (Çolakoğlu, 1991, p. 300).

14 Associations such as Teachers' Union (*Muallimler Birliği*) and Reserve Officers Mutual Aid Society (*İhtiyat Zabitleri Teavün Cemiyeti*).

15 I borrow the term from Nergis Canefe (2002) who defines a patriotic Turkish nationalism as an ideological founding of an ethno-religiously distinct Turkish nation in its homeland Anatolia. According to Canefe, it denies the late Ottoman roots shaping its own core ideas and rejects the historical continuum between Ottoman imperial and Turkish national histories. For Canefe, "there was a tradition of patriotism and communalism bordering on nationalism in the late-Ottoman period, which then led to the formation of and idiom and movement of patriotic Turkish nationalism during and after the Turkish Independence War" (Canefe, 2002, p. 145).

16 Despite Şükrü's reminiscence that his grandfather bought the house in 1926, it should be noted the Law Article 1064 declaring the denaturalization of former citizens convicted of treason and the confiscation of their properties was issued in 1927 (See Beşe, 2012).

17 The British troops occupied the town on December 6, 1918 and handed down control to the French in October 1919 as foreseen in the Skyes-Picot agreement of 1916 concerning the sharing of the Middle East region.

18 The Red Cross Commission cited in Turkish sources is probably the same as what Watenpaugh indicates as the King-Crane Commission assigned with the US president Wilson in 1919 by the duty of gauging local opinion in contested areas where the British and French states have imperialist claims (see Watenpaugh 2005).

19 Several references to the diary can be found in Çolakoğlu (1991).

20 *Şimdi Atatürk bu Kurtuluş Savaşını yaptıktan sonra Cumhuriyet devrinde Ankara'da bir liste yapmışlar. Orada sade Kilis'ten değil, İstanbul'dan, Ankara'dan, Anadolu'da vatana hıyanet yapan her yerden yüz ellilikler içinde var. Ben hatta birini gördüm. Harpten sonra geldi. Hududeli gazetesini çıkarırdı burada. Onunla konuştum. O hıyanet yapmadığını [söyledi]. Tabii o zaman çekişme var ya. Ondan dolayı mallarına el koymak için güya kendisini buraya gelen kumandanlar onu sürdürmüşler.*

References

Beşe, İ. (2009) İşgalden-kurtuluşa Kilis: Aralık 1918–1921, Ankara: Via Design.

Beşe, İ. (2012) "Cumhuriyet Tarihimizde Siyasi Yargılamalar ve Sonuçları", *Kilis Postası*, 17 December. Retrieved from www.kilispostasi.com/kose-yazisi/cumhuriyet-tarihimizde-siyasi-yargilamalar-ve-sonuclari-ecz-ibrahim-bese/1200180; downloaded on 19 February 2014.

Bouchair, N. (1986) The Merchant and Moneylending Class of Syria Under the French Mandate, 1920–1946, unpublished PhD thesis, Georgetown University.

Canefe, N. (2002) "Turkish Nationalism and Ethno-Symbolic Analysis: The Rules of Exception", *Nations and Nationalism*, 8(2): 133–155.

Çolakoğlu, Ş. (1991) *Kilis Direniş – Kurtuluş ve Sonrası 1918–1921–1930*, Ankara: Feryal Matbaacılık.

Durakbaşa, A., Karadağ, M. and Özsan, G. (2008) "Türkiye'de Taşra Burjuvazisinin Oluşum Sürecinde Yerel Eşrafın Rolü ve Taşra Kentlerinde Orta Sınıflar", TÜBİTAK Project No: 105K174, Muğla Sıtkı Koçman Üniversitesi.

Eroğlu, C., Babuçoğlu, M. and Köçer, M. (2012) *Osmanlı Vilayet Salnamelerinde Halep*, Ankara: Orsam.

Gelvin, J. L. (2006) "The 'Politics of Notables' Forty Years After", *Middle East Studies Association Bulletin*, 40(1): 19–29.

Gülcü, E. (2012) "Milli Mücadele Döneminde Kilis", *Kilis 7 Aralık Üniversitesi Sosyal Bilimler Dergisi*, 2(3): 1–37.

Hourani, A. (1993) "Ottoman Reform and the Politics of Notables", pp. 83–110 in A. Hourani, P. S. Khoury and M. C. Wilson (eds.), *The Modern Middle East Reader*, Berkeley: University of California Press.

İnalcık, H. (1991) "The Emergence of Big Farms, Ciftliks: State, Landlords, and Tenants", pp. 17–36 in Ç. Keyder and F. Tabak (eds.), *Landholding and Commercial Agriculture in the Middle East*, Albany, NY: State University of New York.

İnce, H. İ. (2004) *Milli Mücadele'de Kilis*, unpublished MA thesis, Gaziantep: Gaziantep Üniversitesi.

Karadağ, M. (2005) *Class, Gender and Reproduction: Exploration of Change in a Turkish City*, unpublished PhD thesis, University of Essex.

Kevorkian, R. (2011) *The Armenian Genocide: A Complete History*, New York: I. B. Tauris.

Keyder, Ç. (1987) State and Class in Turkey: A Study in Capitalist Development, London: Verso.

Keyder, Ç. (1991) "Introduction: Large-Scale Commercial Agriculture in the Ottoman Empire?", pp. 1–16 in Ç. Keyder and F. Tabak (eds.), *Landholding and Commercial Agriculture in the Middle East*, Albany, NY: State University of New York.

Keyder, Ç. (1999) "İmparatorluktan Cumhuriyete Geçişte Kayıp Burjuvazi Aranıyor", *Toplumsal Tarih*, 68, pp. 4–11.

Khoury Philip, S. (1990) "The Urban Notables Paradigm revisited", *Revue du monde musulman et de la Méditerranée*, 55–56: 215–230.

Kıvrım, İ. (2008) "Kilis ve A'zaz Voyvodası Daltaban-zâde Mehmed Ali Paşa ve Muhallefâtı", *Ankara Üniversitesi Osmanlı Tarihi Araştırma ve Uygulama Merkezi Dergisi*, 24: 147–174.

Köymen, O. (2009) "Kapitalizm ve Köylülük: Ağalar – Üretenler – Patronlar", *Mülkiye*, 35(262): 25–39.

Mardin, Ş. (1967) "Historical Determinants of Stratification: Social Class and Class Consciousness in the Ottoman Empire", *Siyasal Bilgiler Fakültesi Dergisi* 22(4): 111–142.

Masters, B. (2010) "The Political Economy of Aleppo in an Age of Ottoman Reform, *Journal of the Economic and Social History of the Orient*, 53: 290–316

Meeker, M. (2002) *A Nation of Empire: The Ottoman Legacy of Turkish Modernity*, Berkeley: University of California Press.

Mizrahi, J. D. (2003) "Un « Nationalisme De La Frontière »: Bandes Armées Et Sociabilités Politiques Sur La Frontière Turco-Syrienne Au Début Des Années 1920," *Vingtième Siècle. Revue d'histoire*, 78, 19–34.

Öztürk, M. (2005) "İzziye Kazasının Kuruluşu ve Milli Mücadeledeki Yeri", *Ankara Üniversitesi Dil ve Tarih-Coğrafya Fakültesi Tarih Araştımaları Dergisi*, 37: 29–45.

Roded, R. (1986) "The Syrian Urban Notables", *Journal of the Israeli Orient Society: Asian and African Studies*, 20(3): 375–384.

Şahin, A. (1999) "Giriş", pp. 1–33 in *Kilis'li Abdullah Sermest Tazebay Divanı: Metin-İnceleme*, Uygur Tazebay.

Sayarı, S. (2011) "Clientelism and Patronage in Turkish Politics and Society", pp. 81–94 in B. Toprak and F. Birtek (eds.), *The Post Modern Abyss and the New Politics of Islam: Assabiyah Revisited Essays in Honor of Şerif Mardin*, Faruk, Istanbul: Bilgi University Press.

Timurtaş, K. (Kilisli Kadri) (1932) *Kilis Tarihi*, published by O. Vehbi, İstanbul: Bürhaneddin Matbaası.

Watenpaugh, K. (2003) "Middle-Class Modernity and the Persistence of the Politics of Notables in Inter-War Syria", *International Journal of Middle East Studies*, 35(2): 257–286.

Watenpaugh, K. (2005) "Cleansing the Cosmopolitan City: Historicism, Journalism and the Arab Nation in the Post-Ottoman Eastern Mediterranean", *Social History*, 30(1): 1–24.

Yazıksız, N. A. (Kilisli Balhasanoğlu Necip Asım) (1927) "Kilis Ocaklıları ile İki Hasbihal", *Türk Yurdu*, 5(28): 352–359.

Zhurzhenko, T. (2011) "Borders and Memory", pp. 63–84 in D. Wastl-Walter (ed.), *The Ashgate Research Companion to Border Studies*, Surrey: Ashgate.

3 Fall from grace

The decline of traditional landed notables

This chapter aims to understand in what ways border induces changes that lead notable families to experience downward mobility. It reveals that the traditional landed notables, rather than trade notables, suffer from the sense of falling from grace. "Falling from grace" is used in Katherine S. Newman's study on downward mobility with regard to the experiences of the American middle class, who fail in their commitment to the American dream (1999). But it is not a mere figure of speech. In fact, Newman invites the reader to consider that downward mobility is a hidden dimension of American society because it does not fit into their cultural universe. So, the middle class descending the social ladder not only has to cope with economic hardship but also with falling from grace; that is, "losing their proper place." Downward mobility affects the perceptions and values of the individuals experiencing it and these alterations extend over to the broader social net of family.

This chapter discusses the downward mobility experienced by local notables in terms of their feeling of dislocation that pushes them to embrace values and practices, which they once found to be disgraceful. The fact that wealth creation through illegal means is normal and is even aspirational among local notables is illustrative. Capital accumulation through illegal means in Kilis as of the 1960s is the significant factor that leads to the experience of falling from grace among traditional landed notables and distinguishes them from their counterparts elsewhere. Finally, the chapter explores the ways in which their sense of falling from grace culminates in a nostalgic attachment to a home place, which is imagined as part of the city of Aleppo.

Struggling to maintain status distinctions

The division between traditional landed and trade notables does not only consist of a political rift. It is also accompanied by a symbolic struggle in the maintaining of status distinctions. The traditional landed notables call upon lineage and patrimonial heritage in order to display their status. As the recollections of interviewees illustrate, traditional landed notable families also engage in redrawing boundaries with trade notables in reference to culture and consumption, though the trade notables could compete better. This competition among the notables would be the precursor of a broader cultural transformation associated with the money economy rather than paternalistic domination in the 1960s, which will be addressed in the following chapters on the rise of new wealth.

Lineage is a distinctive source of prestige for the traditional landed notables. Members of these families can retrace their genealogy to at least a span of five generations and mention the accomplishments of their great grandfather. These families generally owe respect and recognition to the philanthropy of their ancestor. As large landholders, these families had to rely on paternalistic domination over the laborers working on their farm (Karadağ, 2009, p. 537). The traditional organization of agricultural production required hard work both for farming and domestic chores. The perpetuation of a hidden domination was needed to make poor peasants work for the landlord as sharecropper or renter on the land or as domestic servants, which meant a scanty livelihood for these poor peasants. Thus, the landlord provided care and protected their needs in return for their submission. Such paternalism usually included charitable activities at the community level as well. The landlord, particularly local governors, built mosques, baths, bazaars, and foundations for the town.

The interviews reveal that land was the major source of wealth for these families. Large landholding could require the head of the family to closely supervise farming. Therefore, there was a second mansion on the farm, other than the one in the town. While the farm mansion would temporarily host the family, the urban residence was permanent. Some of my interviewees still stay in their urban mansions. These are two-story stone buildings with courtyards (*havş*), made by artisans skillful in stonemasonry and they are typical examples of residential architecture of the Ottoman Aleppo.

According to the interviewees' recollections, these mansions accommodated an inner division between female and male spaces. In big mansions, the living space was composed of several rooms, together with the kitchen and toilet surrounding the courtyard. This space, called *haremlik*, was for women, children, and female guests, and familial living when there was no presence of other women. *Selamlık* was an extension but a separate space reserved for the male members of the household, where they ate and received their guests. In some houses, a revolving cupboard between kitchen and selamlık helped to serve the meal to the male guests and provided privacy to both gendered spaces. Some heads of households were more prominent, so their rooms received distinguished guests including local administrators, high officials, and members of notables where the audience could exchange conversations about current issues on the local agenda. Having a room, *oda sahibi* as it is called, was a significant sign of notable status and provided access to clientelist networks.

Trade notables who became apparent at the turn of the twentieth century also called themselves old families of the town. But unlike the traditional landed notables, their family histories usually started with their grandfather who was wealthy enough to buy the mansion that would be their natal house. A comparative study on the relationship between housing and mobility in Britain and France suggests that housing is a pre-eminent symbol of status and identity for English aristocratic families and the newly wealthy family could thus view a country house as an asset for upward mobility to the ranks of nobility (Bertaux-Wiame and Thompson, 1997, pp. 134–135). For instance, the recollections of two sisters, Behire and Tijen, whose uncle was a known gold trader in the 1950s, illustrate such property transfer between traditional land notables and merchant families. Their relative had bought the old eye hospital building, run by the Armenian community until their deportation, from a traditional landed family that had put it into use as a dwelling house.[1] Still, paternalistic relationships and the traditional habitus of old wealth could thwart the attempts of trade notables to convert their economic capital into social capital (Karadağ, 2005, p. 23). The trade notables lacked the sense of deep-rooted attachment prevailing among the traditional landed notables for their parental house. Most of these mansions are demolished by the succeeding generations themselves in the rush to convert old houses into modern apartments, stores, and business offices. Others are left to slow degradation.

The housing in Kilis is composed of a continuous and dense architectural structure with narrow streets and stone-laid, barrel-vaulted passages (*kabaltı*), surrounding the block of the dwelling. The typical architectural feature of these houses was their isolation from their surroundings (Bebekoğlu and Tektuna, 2008). Thus, a house located at the end of blind alleys was more valuable. The outer walls reached up to 2.5 meters to detach the interior from the street and if the façades directly faced the street, no window was open on either floor (*tabaka*). Notable families sustained a hierarchy of prestige not only with their economic power but also with a distinctive lifestyle and tastes considered to be 'modern' since the end of the nineteenth century (Karadağ, 2005, p. 21–22). The growth of modernization at the end of the nineteenth century made the adaptation of Western lifestyles and tastes a field of competition between traditional landed and trade notables and its impact is first seen on the changes in the inner space of the home. The opening of regular or oriel windows with latticed screens on the upper floor is illustrative of this impact.

The trade notables could welcome the changes transforming the sphere of privacy even more quickly. The recollections of interviewees show that although they lived in traditional joint households, the gender-segregated division of space had blurred with the transformation of haremlik and selamlık respectively into the living room and the reception room. Together with the wave of urban development in the 1960s, they could even modify the old mansion into an apartment house. The forenamed house that was confiscated by the local authorities when the proprietor was convicted by the Independence Tribunal and bought by a trade notable family is illustrative. During my visit to the mansion in order to interview Şükrü, the heir to the mansion who was residing there, he showed me around. Şükrü's father had made several modifications to accommodate two families, father and son, under the same roof and modified the two-floor mansion into two flats. As Şükrü detailed what modifications were made, some were massive changes to the original features. For example, the old floor covering, made by a traditional building material called *kursümbül*,[2] was replaced by cement and the old indoor marble stairs leading to the middle of the downstairs hall (*sofa*) was supplanted by a narrow cement stairway at one side in order to cut off the upper floor from below.

Traditional landed or trade notables, women played a particular role in transmitting new cultural values into everyday life. They were exemplary in representing the proper ways of being 'modern' which is associated with the body, social class, and consumption patterns and thus, they became arbiters of social distinction (Karadağ, 2009, p. 543). Although the public space is highly segregated on the basis of gender in the early Republican period in the provincial cities compared to the centers, the changing fashions of clothing and interior decoration reflect the shift in tastes. The memories mentioned by notable women themselves illustrate how they were eager to keep up with the changing clothing fashion.

During the early Republican period, these women used to order fabric from Istanbul and Aleppo and shop for clothing and shoes from Aleppo, a city under the French Mandate that could provide access to Western goods. An important sign for the refashioning of taste was the realm of consumption for displaying lifestyle. An example would be the demand for decorative furniture. For instance, full-length and framed decorative mirrors with elegant motifs of birds or leaves carved on solid wood came into vogue, as mentioned by several interviewees.[3] In particular, mirrors with a trough at the bottom—called, mirror with jardiniere (*bahçeli ayna*)—were the most favored.

The recollections of interviewees emphasize that at the first decades of the new Republic, as young girls and married women, they were not yet allowed to go shopping and it was the duty of the head of the household to purchase the daily food and send it home with the help of a male servant. They were seldom allowed trips to Aleppo with the company of a male member of the household and to visit relatives. But their greater participation in social life contributed to the refashioning of lifestyles and tastes. For Latife, born in 1911, who says that her family descends from Kırım aristocrats granted land in Kilis by the Ottoman government, the fashion of clothing in the town indicated a state of being civilized, mostly associated with urbane manners of living and opposed to the rural. She tells how she was received by the notable families of Bursa in 1942, the year she went as a newly married young woman, as follows: "I came to Bursa as a newlywed. Now imagine that they think of Kilis as a village and my entire dowry was tailored and came from Aleppo. They were so stylish."[4]

They revealed visiting days (*kabul günleri*), outdoor social gatherings, or refashioning of traditional rites into new forms such as cinema weddings as utterly modern practices. The adoption of these modern practices among the notable women could help to reproduce social cohesion among their stratum and extend class boundaries to other social segments (Aswad, 1974, p. 10). As I was invited on a few visiting days organized by notable women, I observed the intermarriage relations connecting families to each other was a prominent factor behind the social networking among notables, though the notable women could also develop social contact with their middle class peers. Memories of gathering habits illustrate how these social occasions allowed women to reproduce boundaries of distinction. Interviewees' recollections implicated how parks could transform into the iconic spaces of modernity in the provincial localities,[5] where status distinctions could be performed publicly:

They did not serve tea or coffee in Ayşecik Park[6] before. You could just go and sit there. The place of notables and the place of normal families are separate. I mean, they adopted this opinion as such. They used to do it automatically. In fact, there was no one who cherished them particularly. They thought themselves as superiors, as if we are not worthy of them. They think they cannot sit with us and they go and sit at place they are worthy of. But there is no one to tell them, "you are from such a good family, so you should sit there" really. They just accustomed themselves to do it at a young age.[7]

The shift in lifestyles and tastes also became apparent in the transformation of urban spaces. In the late 1960s, the number of cinemas was seven and they welcomed theatre plays, concerts by vocal artists at the harvest seasons, and indoor and outdoor movie screenings. They made new socio-economic strata visible as well. The notable families were denizens of foreign movies in City Cinema, also known as *Ebe Hanımın sineması* after the woman who was the daughter of a traditional landed family and whose education was midwifery. On the other hand, middle and lower class women could also go to watch Turkish movies either in the remaining big hall of the old demolished church reconstructed as a movie theatre or elsewhere. The old hall also held "weddings in cinema" that came into fashion later. The guests would be entertained during the wedding by watching a movie scheduled by the cinema after the bridal ceremony.

Beyond the border: *eşraf* losing ground

Although the land was a lucrative estate, wealth creation among the landed notables did not originate mainly in large-scale agriculture during the early Republican era. The increase of trade did not alter the organization of agricultural production but it would afford the landed notables the chance of diversifying their revenue-bringing ventures and provide access to the trade markets in the post-Ottoman Greater Syria where they continued to sell their crop. Interviewees' recollections from as early as the 1940s reveal the illegal trade of gold and other products across the newly demarcated border with Syria. I will trace the shift of internal regional trade with Aleppo into a cross-border trade through these interviews, as well as through official documents. I assume that exploring the development of cross-border trade would reveal social mobility strategies among the traditional landed notables. Cross-border trade promoted the trade notables and the traditional landed notables' started to lose ground to them. The trade notables could quickly adapt to the situation due to the intergenerational occupation transmission. That is, they had already learnt how to trade from their fathers. Still, as I elaborate on the patterns of cross-border movement, I will be able to portray the possible ways that the traditional landed families, though limited in number, could overcome the traditional barriers against their conversion to trade.

Also the traditional landed notables literally lost their landed estates at the Syrian side when their properties were expropriated by the government in the 1960s within the context of agrarian reform. They define it as a major relapse in their economic prosperity. The development of cooperative relations with Syria recently raised the prospects of the traditional landed notables reclaiming the properties left in Syria after 40 years of struggle. The families took legal action while they also addressed government with their petition to recover their landed estates confiscated by the Syrian government in the 1960s.

Passavant gates, illegal trade and the 'golden' opportunities of border

The demarcation of the international border and the break up from French Mandate Syria constitutes a major shift in the internal regional trade between Kilis and Aleppo. Before the delineation of the border, trade in Aleppo was based on the importing of European manufactured goods and the exporting of agricultural commodities. Locally produced merchandise from the handicraft industry, as well as sheep, cattle, wool, and butter brought from southern Anatolia constituted important items of commerce (Bouchair, 1986). The dislocation of southern Anatolia meant the loss of a large trade area for Aleppo. It also marked a decline in traditional industry, together with further European economic penetration with the Mandate regime. However, the interviewees' recollections demonstrate that the regional trade persisted despite the border demarcation, turning it into an illegal trade that the state authorities were helplessly trying to regulate.

The state authorities were concerned to take preventive measures against the "smuggling" practices by indicating the unauthorized and uncontrolled movement of goods at the border gates as early as 1931. The year is marked by one of the renowned reports by Şükrü Kaya, the minister of internal affairs, upon his inspection tour to southern and eastern provinces of the newly founded nation-state. As he recorded, contraband trade grew in the last couple of years to the degree of compensating Aleppo's loss of its northern and eastern hinterland with the border and almost reviving the city as a trade and industrial center (BCA, 1931). Kilis received the smuggled goods of salt, flint, silk fabric, gas oil, and sugar, with oil and sugar having the highest percentage of contraband in the town.

The minister's report also stated the significance of border gates that served as the access points for proprietors and peasants to work on the landed estates at the Syrian side. A specific transit regime with a document, called *passavant*, was agreed upon by two countries[8] to regulate the border crossings of landholders with their laborers as well as their pack animals and equipment required for farming. According to Kaya, the number of daily crossers could amount to 5,000 people when the fields would be ploughed or reaped. As the report would illustrate, the demarcation of the border meant that the landed estates of most notable families were left on Syrian soil. The travel permit allowed the household heads, as the landholder, cross-border movement in order to gather their harvest. In addition, it generated for the landholder the condition of circumventing the passavant regulations for yielding extra revenues of trade until the unilateral decision of Turkey to close the passavant gates occurred in late 1966.

Family members frequently indicated that they continued to visit their farms at the countryside, often extending their trip to the city of Aleppo though they needed to issue a passport and visa for such long distances. The passavant gates helped the landholding notables abuse their rights as proprietors and cross the border in two ways: by manipulating passavant regulations to sell their crop at Aleppo or by buying up some goods from Syrian markets to smuggle back into the country. Although as late as 1966, the parliamentary question raised by the deputy of Hatay regarding the circumvention of passavant regulations is illustrative of its nature. The passavant agreement permitted the landholders domiciled within 5 km of either side of the border to cross the border without declaring the harvested crop at customs. Thus, the deputy assumes that if the crop is sold at the foreign market, the transit agreement would be misapplied and the price difference between two countries would generate extra gains for the landholder-trader.[9]

On the other hand, the landholders could also abuse it by writing up the total market value of the their crop and buying the fictitious amount from the Syrian markets at longer distances than prescribed in order to cross them to the Turkish side without customs clearance. The deputy of Hatay would poignantly indicate in the parliamentary question the smuggling at the border as one of the gravest ills eating into the Turkish economy:

> Now, [the landholders] have to apply to the Passavant Commission in proportion to the amount of cultivated land according to the available Passavant Agreement. Then there will be a committee for setting the estimation. This committee will go and make a rough estimation. It will set 300 [kg.] of cotton instead of 100 per hectares and four or five times more for the wheat where it is not possible to harvest 20 to 40 kilos per hectares. In terms of paddy, it will set four or five times more and therefore, despite the fractional amount yielded, [the landholders] will buy crop from the market and here and there and they will bring Turkish goods to Syria, whereby they will enormously engage in smuggling. My dear friend, this smuggling affair is not something to be neglected as it climbs over billions [of Turkish liras].[10]

Even though it is not possible to gather data about which families among notables were involved in contraband, the interviewees' recollections suggest how they were able to yield high gains from their landed estates. Female members of the family do not have full knowledge of their father's or husband's running of affairs but all of them said that they used to earn a good deal of money from the sales of produce. The families often had a deputy (*vekil*), responsible for watching over the farming and selling of produce, and sent the yield to the family.

The illegal gold trade was another yet subtler means of yielding extra benefits from cross-border movement. Within the framework of peripheral transformation in the late Ottoman era, the increased trade did not yield the market-oriented agricultural production on the landlord-managed estates, but "it did allow various well-placed officials to benefit from new opportunities in the circulation of products and money" (Keyder, 1991, p. 5). This argument could be extended to the early Republican economy at the border, where certain members of notables could take advantage of their social and economic power. Needless to say, the gold trade was strictly regulated by the state until its trade has been liberalized by the neoliberal governments of the 1980s.

Until the end of World War II, the farmer notables used to sell their produce in exchange for old Turkish silver coins and gold pounds. The Ottoman currency system in Syria was kept in place after the break out, especially after the devaluation of the French franc in 1926 (Bouchair, 1986). With the loss of value in silver coins in 1935, Aleppian merchants demanded their payment solely in gold, which probably meant augmentation in the volume of gold circulation. The interviews reveal the smuggling of gold in bullions was already in place in the late 1930s. Asuman, a woman of trade notables born in 1922, remembers that her father put the smuggled gold out of sight by hiding the gold in her clothes when she was a child. She used to wear a waistcoat tailored for that purpose with pockets on her back for placing gold bullions under her casual cloths:

> Well, they [my parents] made me to go to Aleppo. My parents used to go. I was little, about seven or eight years old. The officers liked me. They used to come and check [the cartridge]. My father told me to wear it; they would make me wear a kind of smock. On the smock, they would all line up the gold bars. They [the officers] would not search me. However, there were bars under my shirt. The officers would take me on their lap and kiss and let me go. They would not meddle. Indeed, they would not let anything to pass at some other time. Since they [the officers] liked me, my late father, he used to line them on my back. You see, I was little and I remember it.[11]

The gold trade generated wealth and prospects for gentrification for the trade notables. Behire and Tijen, born respectively in 1933 and 1940, said that their uncle was able to buy the old eye hospital building, left over from the Armenian population, from a traditional landed family. The way they depicted the old stone building with two courtyards and fountains suggests that the house was exceeding the mere function of dwelling and it would be a source of prestige and admiration. The interviews also hint at the ways in which traditional landed families benefited from the unobtrusive circulation of gold across the border. Among the interviewees, many remember their father and husband's frequent travel to Aleppo, Damascus, and Beirut. But as Hikmet, an elderly woman of traditional landed notables above her eighties, recalls it was not the male members of the household but their men, working for their patrons, who carried the smuggled gold.

"Dear Prime Minister": negotiating for landed estates left at the Syrian side

The interviewees' recollections and the heated debates in the parliament illustrate how much the loss of these properties had been significant to the traditional landed notables. My discussion in this section follows the long history of this dispute of land with Syria from the viewpoint of traditional landed notables. I assert that traditional landed notables shifted to adopting the victimizing language of border citizens in order to negotiate their claim over their properties in Syria, rather than negotiating their *eşraf* privileges by calling upon their lineage and patrimony.

Until the start of the Syrian civil war, the then Prime Minister Erdoğan was seen as the architect of peaceful relations between the two countries and was expected to provide the solution for this long-standing dispute. An open letter by a member of the traditional landed notables addressing the Prime Minister shows that the anticipations of "real estate victims" had been excited by his Syrian diplomacy. Bahadır, a member of traditional landed family born in 1956 and proprietor of landed estate in Syria, had written an open letter addressed to the Prime Minister that said "real estate victims were expecting him to herald them the good news" in early 2010. Though he gave me detailed information, he also provided me with a copy of his letter. There he put the case with reference to the long history of this messy problem. The lands owned by Turkish citizens were disputatiously confiscated by the Syrian government in the mid-1960s. When the Turkish government reciprocated by seizing Syrian properties in Turkey, the Syrian government compensated their citizens by redistributing the properties confiscated from Turkish families. But the Turkish citizens who used to own properties in Syria could not recover them. Their efforts proved inconclusive until the issue was retaken by the AKP (*Adalet ve Kalkınma Partisi*, Justice and Development Party) government to settle the mutual exchange of confiscated properties. As Bahadır wrote, it was to the knowledge of families that about half of two billion hectares alleged by Turkish citizens as their own properties were covered by the settlement of an international agreement with Syria.

While I have been conducting the interviews with local traditional landed notables, the gradual increase in popular unrest in Syria pushed the prospects for exchange to uncertainty. According to hearsay, the exchange of properties between the two states was agreed, but the negotiations were not completed when civilian unrest broke out in Syria. The hopes of interviewees for the exchange were already broken when I listened to them. Still, I could observe how much the recent negotiations had raised the expectations of boosting the 40-year long struggle of the families. As rightful proprietors, they could welcome the chances for sudden prosperity once again: they estimated the real estate value of these fertile agricultural lands to be 30 to 50 billion dollars in total. Bahadır indicates that 3,800 families were victims of property loss.

My interviewee Ayhan, a woman from traditional landed notables, hoped to receive at least a share from the rent incomes collected by the Turkish General Directorate of National Real Estate in exchange of her confiscated property in Syria. She believed that the rightful proprietors as old landholders who even paid a tithe (*aşar*) to the Syrian government for their agricultural produce should be entitled to have their share of the rent income.[12] The confiscated properties of Syrian citizens in Turkey have been rented by the General Directorate to the farmers since 1966 and collected in a bank account in their name. But the Turkish government refuses to distribute the rent income among the Turkish families in order not to violate the rights of Syrian proprietors.

The Syrian government relied on the 1958 Land Act for the confiscation of the landed estates owned by Turkish citizens. But the actual seizure of the land took several years. Nevertheless, the recollections of interviewees define it as an abrupt happening which they did not expect and had a negative impact on the prosperity of their family. Apart from their real estate value, these lands were a source of income yielded from agricultural produce. Most traditional landed families had their men in Syria looking after the farm, harvesting and selling it on the market in the name of the landholder. The detailed and rather dramatic story told by Hikmet is exemplary in indicating how the news about the loss of landed estates was received by traditional landed families. Hikmet, as a daughter of a rooted and reputed Ottoman dynasty family, almost mourned the loss:

> We had a helper called Hamparsum. Since we had our olive grove in Syria, he looked after it and brought the yield. One day he came to give the news. I opened the door; I had my flat on the top of İşbank building. He said: "Come my girl". I went downstairs. In the past, we could not speak so freely to our male acquaintances. I said: "Go ahead effendi". He said: "You might send your step son. Tomorrow, I will give your money. Let him come and take". He had found a buyer for the olive harvest. ...Now, we left it over there that day. But I was delighted. We would receive some 6,000 liras and 6,000 liras was big money forty years ago. Oh my! I was so delighted. The kids were little and we would have a sizeable amount of money. It became evening. My mother-in-law and my sister-in-law were with us. We all went to my parents to pay a visit in the evening. Here, Raika [her helper] was also with us. She made coffee for us. We were sitting. I cannot forget that day. I remembered it recently, the other day I lost my sleep over it. Well, I had the coffee noggin in my hand and I was drinking it so heartily. Just then, my elder brother came and came upstairs. Everyone's fortune is told. Mine [the noggin] was still in my hand. He said, "Did you hear the news?". They replied, "What?". We all had our estates there. He said, "Today, the Syrian government seized them. It is over". I mean that day, all of a sudden. I mean, I cannot forget that day. I told to myself, alas, I am dead. I flipped my hand unwittingly. I should have flipped it hard because of the pain. Somehow, my ten fingers were all black the other day. All that property and 6,000 liras had gone. We could not get the money back either. Life has gone by just like that.[13]

The interviewees mention about the recent efforts to reclaim their properties but few have memories about the reactions to the outcomes of Land Reform by the Syrian government in the 1960s. In fact, Syrian Land Reform had been a matter of debate addressed by the parliamentary discussions in the early 1950s. A member of the local traditional landed notables, elected as Gaziantep deputy, brought the issue of a land dispute with Syria to the agenda of the National Assembly as early as 1952 (TBMM, 1952, pp. 122–127). Though his main concern was the difficulties that the landholders were having over the exercising of their full rights of proprietorship in Syria, it is understood from his speech that Syrian land reform was already revealed as a potential threat. He raised the question of whether the Syrian government had enacted a law for the redistribution of land to the peasants, if the law was binding for the Turkish landholders as well and what diplomatic attempts were made by the government. The Minister of Foreign Affairs would reply that the Syrian government was planning to distribute the state-owned lands and Turkish landholders would be better to renew their registry at the Syrian land offices as an added precaution. These debates were heated anew in the mid-1960s after the reciprocation of the Turkish government with the confiscation of Syrian properties in Turkey and the unilateral closing of Passavant gates.

In particular, the confiscation of Syrian properties in Turkey was the target of controversial debates accusing the government of redundant retaliation against the Syrian government seizing few Turkish properties. The debates were culminated in a motion of no confidence introduced by the Republican People's Party (RPP) in opposition and supported by other opposition parties against the Justice Party government in 1968 (TBMM, 1968a, p. 17). The deputies proposing the motion argued that the Syrian government set a limit of size regarding the lands owned by foreigners within the context of land reform and nationalized only the surmounting amount of one's estate. Thus, a limited number of families had been affected. However, the Turkish government confiscated all properties of Syrian citizens in Turkey without investigating the matter thoroughly and provoked the Syrian government into enlarging the scope of land reform. A short time before the motion, a member of the traditional landed notables from Kilis, elected as RPP Gaziantep deputy for the Republican Senate, had posed the same question to the government as well as criticizing the closing of the Passavant gates for retaliation purposes (TBMM, 1968b, p. 1012). The Justice Party government rejected the allegations by stating that the number of Turkish citizens filing a complaint for their landed estates in Syria accrued to such a level that the confiscation of all Syrian properties was not enough to cover the equivalent number. In all these debates, the deputies had underlined the victimhood of border citizens who had to earn their living from their lands at the other side of the border.

The notables of Kilis expected that as the government diverged from the old politics and initiated cooperative policies, a compensation for the victims such as the exchange of confiscated properties between the two states could be probable. Unfortunately, the relapse of Turkish international politics with Syria into hostility again failed these expectations and added to the experience of traditional landed notables of falling from grace.

"We lived like a cicada and they worked like ants": new encounters between *eşraf* and *esnaf*

The start of disintegration of large-scale landholding in the 1950s had already forced the traditional landed notables to adapt to new conditions of living. Hence, they had to assume new strategies of upward mobility in order to secure their social position. Ironically enough, the social reproduction of their class identity as traditional landed notables required them to cross over the boundaries of distinction with the trade notables and embrace an alliance with them rather than competing with them. Lacking any occupational heredity, the traditional landed notable families could only safeguard their position by reconverting their capital held in one form into another. Investing in trade, urban employment, out-migration, and marriage are the principal means of social reproduction for traditional landed notables.

Even though traditional landed notables maintained middle and upper class positions among the native Kilis community, they experienced the cost of change as downward mobility. The story of Ferit, a member of the traditional landed notables, illustrates the experience of "falling from grace" among their stratum as the boundaries between *eşraf* and trade *esnaf* fade away, inducing the former to embrace what was once disgraceful for them. Him telling his life story with reference to the famous fable of Aesop about the cicada and the ant signifies his regret about the delusion that traditional landed notables' privileges would be everlasting and his admiration of the new wealth's success in upward mobility.

I interviewed Ferit at his office in a dentist clinic where he works. He squeezed our meeting in between his consultations. As he remarked, he had to continue working at this age—he was born in 1939—and he ascribed the reason to his shortsightedness in planning his future. He supported his statement by showing me the pictures hanging on the wall of his office. The pictures were the stills of a movie shoot in 1965 showing Ferit on his own Harley Davidson motorcycle as part of a rebel youth gang with some movie stars. His aspiration to appear in movies made him quit his medicine education in Ankara, which he regrets a lot today. Later, he could only advance in dentistry by being re-enrolled for a university education in Istanbul. Ferit exemplified the decline of the traditional landed families with the disintegration of large landholding in the countryside, which was addressed by the Turkish film industry with the caricatured figures of spendthrift heirs of wealthy notables. His story also illustrates how the members of traditional landed notables eventually had to reconvert their economic power to social capital by getting a university education in order to obtain urban employment to safeguard their position.

The traditional landed notables are the *rentier* class. Unlike the trade notables, they did not have occupational heredity and turning themselves into traders without the necessary skills and aptitude was difficult. The trade notables could take advantage of intergenerational occupation transmission by handing down "the family business" to offspring. That is why my interviewees among the merchant notables described their lineage as *esnaf* family. The traditional landed notables seem to benefit from the opportunities offered by the shift in regional economy into an illegal border trade as additional income. It is only with the start of disintegration of large-scale handholding in the 1950s that these families were forced to adapt themselves to new conditions of living. As Ferit remembers, he and his friends found odd that one of their fellows started to work, as they were not used to it.

For early generations, education was the symbolic capital required for sustaining paternalistic relations. Children of traditional landed notables used to receive their primary education at local madrasahs. The junior high school was opened in 1915 and the senior high school was established only as late as 1958 in Kilis. The young generation had to move out of the town after secondary education and receive high school education elsewhere as in the case of Ferit who graduated from Kabataş High School in Istanbul. The early generations used to receive university education abroad. There were a few Sorbonne graduates among the leading families, who wanted to be *au courant* with modern mores. Especially for members of ulama families, their role as teachers enhanced their influence over the local community by associating religion and education. Local madrasahs were educational establishments that combined the teaching of the Islamic canon with lessons on grammar, letters, inheritance law, and logic. In other words, their interest in education secured symbolic profits for landowner notables (Karadağ, 2005, p. 146). Nevertheless, agricultural income would barely sustain the expenses that came with a university education and a separate household in Istanbul or Ankara in the 1960s, as Ferit notes. He recalls that his father had to sell a roomful of cotton in order to afford to pay the school expenses of his elder brother.

Ferit believes that the division of the large lands among their heirs was the principal reason behind economic decline. His remarks resonate with the scholarly literature about land tenure in Turkey. The prospects of traditional landed families for sustaining their class position were closely related to the process of rural change in the 1950s (Karadağ, 2005, p. 113). The agrarian policy of the Democrat Party in power targeting the development of the peasantry, together with rural migration to the cities, caused the collapse of sharecropping arrangements on the land. The US-granted Marshall aids were channeled to the mechanization of farming and the importing of tractors. Moreover, the migration to the urban settlements provided opportunities for employment in the labor market, turning the peasants into seasonal and wage-labor. The landholders could not find tenants to make sharecropping arrangements. Furthermore, they also had to deal with growing unrest among their tenants, who tried to reclaim the land.

The fragmentation of land among the heirs as patrimony after the father's death increased the tendency of selling it off. For Ferit, it meant the wasting of large, fertile agricultural lands and vineyards as they were sold piece by piece in order to cover their education and living expenses in the city. This shift in land tenure also marks a decline in the familial meaning of land as status indicator. The Ottoman pattern of inheritance could be contrasted to the English gentry who could retain the patrimony undivided by handing down through the male primogeniture to a single heir (Bertaux-Wiame and Thompson, 1997, p. 133). On the other hand, the Ottoman waqf system aimed to secure the proprietorship of the land as patrimony and its bequeathing to all heirs as inheritance. The pattern of inheritance did not allow the inheritance of all property by one child and divided the land into parcels. The land was not considered as an economic patrimony to be sold, but as a paternal patrimony to be kept.

Ferit's life story dramatically contrasts the old wealth with the new rich and puts emphasis on the new encounters between them. He retrospectively describes their living as one of luxury, even lavish and inconsiderate. His reference to his life story by drawing on the parable of the cicada that enjoyed the day and did not save up anything for later is asserted as the reason for economic decline. His youthful memories of summer holidays as he came back from summer schools revolves around the partying of traditional landed notables' children in their farm houses, riding horses or ostentatious motorcycles. The moral of his life story into the parable of the cicada, however, stems from the rapid ascendancy of new wealth. The members of the traditional landed notables like Ferit witnessed the climbing of the lower strata on the ladder of upward mobility in a shorter time than one's lifetime and without the occupational transmission of the elder generation. For example, he mentioned that the son of the gardener who used to work for his relative succeeded to make enough fortune to start buying lands in Istanbul along the well-trodden highway to the airport in the mid-1960s. The real estate investments of the family make them among the top ranking businesses in the city today.

According to Ferit, wealth accumulation obtained through illegal trade helped the lower strata climb up the social ladder to an unprecedented scale and the traditional landed notables declined. This implied for him a subversion of the social order and the beating of old wealth by the lower strata, as his reminiscence of local hearsay demonstrates. The hearsay goes that the new wealth now boasts of their Mercedes cars while comparing themselves to the traditional landed notables who used to ride their motorcycles in the town. Ferit states that the reason for the decline of traditional landed families is their inability to engage in illegal trade, which they regarded as shameful:

> [It ended like that] because there was the habit to use up what is on hand rather than earning it and it has been divided up as inheritance, and because these gentlemen could not engage in smuggling, as smuggling was a disgrace for them. My father did not let us wear nylon socks. The nylon socks came from Syria and we could not wear it. Although our village and land was just along the borderline and our employees, majordomo and sharecroppers were smuggling, we could say, "Here, have this 100 liras and buy for us too so that our share gets invested". We did not say or rather, we got embarrassed. Therefore, they won ten to one or hundred to one. We could just earn one for hundred and we could not keep up. Later on, those who received education did not turn back to Kilis. He became a doctor and stayed in Istanbul. He became a lawyer and stayed in Istanbul. Or else, he moved to Adana as he was engaged in trade business.[14]

It is worth noting that Ferit's reference to the upward mobility of fellow townsmen is significant in showing the reshuffling of the social insignia of status and how the lower strata, once denied of respect and reputation, achieve them. Although he knows that the new wealth has made their fortune from illegal trade and enterprises, he presents their ascendancy as an achievement by mentioning the parable of the cicada and the ant. For him, traditional landed notables' children used to rejoice in summer holidays with their pocket money from their parents' while their friends in town worked. Ferit says that he had once asked his close friend engaged in the smuggling business why the latter did not let him own a share as well. Thus, the traditional landed notables embraced what they did once consider as shameful and thus, undermined the long-established distinction between traditional landed and trade notables.

The confession made by Hikmet, my interviewee from a traditional landed family, that they used "not to give away their daughters for marriage to a smuggler" points out that the new wealth is now reputed among the traditional landed notables. For instance, Hikmet's son, whom I interviewed as well, is a lawyer who defended smuggling cases prosecuted by the local authorities after the 1980 coup in Kilis. Marriage is the strongest indication of the shift in the traditional landed notables' perception regarding the new wealth, as well as the dissolution of traditional norms and habitus that used to thwart the attempts of new wealth to convert their economic power into social or cultural capital. Under the new circumstances, the role of marriage proves effectual for the traditional landed notables in establishing alliances with trade notables and safeguarding their class position.

Ferit experienced dramatic downward mobility because metropolitan social networks provided new encounters among the members of old and new wealth.[15] Ferit was among the founders of the Kilis Culture Association in Ankara and Istanbul. After a while, the new wealth in Istanbul left to set up another association called the Kilis Foundation. The latter comes to the fore among the benevolent associations of Kilis migrants with broader philanthropic activities such as sponsoring education and the boarding of native students in the cities, and relief for the poor and donations for public school buildings in the native town and has a wider outreach through various uses of media. The foundation, chaired by a prominent *nouveau riche* known for his mafia-esque connections, mostly recruits from the new wealth who had to desert the town as they were criminalized and were sued as smugglers by the military government after the 1980 coup. Members of new wealth had to carry their investments to the cities in order to launder their "illegal businesses" since there was no investment area in the town where big entrepreneurs could not maintain a low profile. Ferit had acquaintances with them.

Reproduction strategies among traditional landed notables consist of conversion to trade and mobility through education and marriage. Also urban migration provides traditional landed families with the opportunities for more rewarding employment and access to the social networks of new wealth and thus, it constitutes part of the reproduction strategies. Despite Ferit's emphasis on the fragmentation of inherited lands, land tenure in Kilis is still dominated by large landholding (Kesici, 1994). Yet, no families among the traditional landed notables are involved in commercial farming. Among my interviewees, only one is engaged in large-scale commercialized agriculture but he is a descendant of trade notables.

For most interviewees, the conversion to trade required a generational leap and the grandchildren of the landholders have now consolidated their investment in trade business. For instance, the nephews of Ayhan, a woman of traditional landed notables born in 1924, run their business in the Grand Bazaar today. Ayhan states that she had gone through financial difficulties after the Syrian government seized her family's landed estates. She complains that many contraband goods were coming to Kilis across the border, but her husband could not do the trade since he was not familiar with it. The traditional landed notables' experience of falling from grace consists in their coping with the reproduction of their class position under new circumstances. As they had to acknowledge the upward mobility of new wealth, they also recognized the respect and reputation that they once denied to them. The illegal cross-border trade as of the 1960s is the significant factor that leads to the subjective experience of decline among traditional landed notables.

"There was no border": nostalgia for the home-place

The large rate of out-migration and the rapid transformation in the town has probably made the questions of roots and belonging more pertinent. Whether left behind or residing out of the town, the elder generation of notables has a more acute nostalgia of the past in contrast to my interviewees from other strata. The interviews reveal the nostalgic memories of past lives that accentuate lineage and patrimony, as previously discussed, as evidences of roots and belonging. I will delve into these memories once again, this time to elaborate on the notables' imagination of hometown as a place of belonging. The interview excerpts in this section show that the notables build their self-identity on the perception that their town has a common social world and refined culture with the city of Aleppo.

The nostalgic experience not only defines the past time as lost and mourned, but it also invokes a longing for home. The word nostalgia, of Greek etymology combining *nostos* (home) and *algos* (pain), originally implies the longing for a lost home and the pain from its irrevocable loss (Malpas, 2012, p. 163). Nostalgia is a stronger disposition "when social change is rapid enough to be detectable in one lifetime; at the same time, there are must be available evidences of the past—artefacts, images and texts—to remind one of how things used to be" (Yeoh and Kong, 2006, p. 57). Kilis town has gone through a rapid transformation with its imprint on the urban landscape in the 1970s, together with the growing urban sprawl around the historic downtown. In a tour around the old town center, it is still possible to locate the remains of old mansions and inns used as industrial workshops for soap or olive oil making. Deteriorated in form or completely demolished, these monuments cannot be further destroyed nowadays due to the recent regulations concerning the cultural preservation of historic sites. The nostalgic return to the past is closely related to the notables' need to identify history as belonging to them and to which they belong.

The members of notable families residing in the town center live in the midst of the ruins of their past legacy. Their ancestry had contributed to the development of the town. The proof of their roots, to which they refer to render their identity as traditional landed notables cogent, is inscribed all over the place. Interviewees mentioned occasionally about the mosques, baths, inns, and foundations that were built by their grandparents. They gave me detailed itineraries for helping me to get there and see on site the places that they were talking about. Their memories recalled the building used by their great grandfather as the government house or as a madrasah. The fact that the family's past is embedded in the urban landscape clearly turns these buildings into a testament of their rootedness and generates an entrenched attachment. Therefore, the town's loosing of its historical identity is likely to enhance the notables' experience of falling from grace, with the corrosion of everything that they have assumed to belong to their heritage and which made their town a home-place.

Perhaps, this is why I observed that a new politics of nostalgia is being been fueled recently among the notables. This is a politics of nostalgia consisting of returning to the past and preserving the cultural legacy of the town, though in an attempt to fabricate the place as a tourist attraction and turning it into a show-case of history with exclusive emphasis on concrete edifices, facades, and visual qualities of historic buildings.[16] A program launched for raising the interregional tourism activities between Turkey and Syria was expected to contribute to the town's development by taking advantage of the increased geographical mobil-ity across the border until it fell into abeyance with the breaking of the popular unrest in Syria and the relapse of Turkish policies of cooperation with the Assad government.

As the interviewees' recollections reveal a yearning for a lost home, they also point to an alternative imagination of place that differs from the hegemonic rep-resentations of the town as a territorially bounded space. The imagined place of belonging in the notables' memories is not a bounded territory within the confines of the state border. But it extends across the border to the farm house they slept in and to the shops and streets of Aleppo they enjoyed visiting. In other words, the imagined home-place continues to be a geographic extension of the city of Aleppo for decades after the demarcation of the state border. Nevertheless, it should be noted that this imagination of place revealed in the notables' nostalgic memories is incommensurable with the contemporary notion of transnational space consti-tuted by cross-border migrations.

Border studies underline the possibility for multiple belongings in the contem-porary age as the means of communication and transport invade the sacred spaces of home and locate them in transnational space (Morley, 2001, p. 432). But the imagined homeplace of the notables cannot be conceived through the angle of transnationality. This notion implies transgression of the national boundaries or of the threshold to a different nationality. In the notables' recollections of the past, the border is simply not perceived as the border as limit, the imagined threshold to another entity. As the notables' plea to the American Red Cross Committee to stay annexed with Syrian Aleppo demonstrates, the border is not only imposed by the state powers, but also negotiated by local actors. The borders are subject to con-tinuous change and negotiation. Even when the borderline is already materialized for the border dwellers, the meanings over the place and relations across the bor-der can be repeatedly redefined (Wilson and Donnan, 1998, p. 21). Virtually, the border did not exist for the notables till they felt the bonds attaching them to the other side of the border were severed, their life world was constrained and their homeplace turned into something that it did not used to be. According to Vefa, a member of the trade notables born in 1943:

[The beginning of illegal trade] happens upon the mid-1950s after the mining of the border zone. After all there was no border among us. There was no border between Kilis and Aleppo.[17]

When I asked him about the demarcation of the border in the early 1920s, he replied as follows:

> No, there was not any [border]. After all, Aleppo was annexed with Kilis; it was an Ottoman province. There was exchange of good and trade business. For example, silken fabrics came up from there, sine satin, enver satin etc. There were lots of name for it. There was no border. They used to bring good from Aleppo to sell them in Kilis and bring goods from Kilis to sell them in Aleppo, just like you go and sell to Antep.[18]

The excerpt from the interview with Vefa illustrates the periodization that most of my interviewees have made in order to draw a distinguishing line. The notables usually emphasize the mining of the frontier zone, in setting the beginning of changes concomitant with the border. For example, a former senate member whom I interviewed in Ankara, assumes that the frontier zone is mined by the Democrat Party government as the consequences of misguided policies toward their town. The notorious labelling of the border town as smugglers' city led the government to take preventive military measures and to mine large pieces of fertile agricultural lands, though the trade activities of local dwellers did not deserve such labeling. As the quoted excerpt exemplifies, the notables called these economic activities "exchange" or trade for subsistence rather than (market-oriented) smuggling. According to them, the illegal practices that could be called smuggling began only after the mining of the border. But it should be noted here that the gold trade is completely out of picture in the view of notables. Thus, with the border, according to the interviewees, the town was transformed into a geographically squeezed and economically curbed place.

Their recollections then reveal nostalgia about the good old days of their town life when there was no border. Their memories cover the convivial meetings of extended family members, often living in joint households or the hosting of other notable families, enjoying the countryside during farm trips, picnics in the gardens and the grape harvest seasons or cooling off during the hot summer around the pond in the courtyard of the house. These memories reveal the innermost feelings about domestic life and also indicate a remembrance of their being, a remembrance about how they used to be. The good old days were the times when the notables felt they prospered. Frequent references to affluence, well-being, and comfort are significant in the favoring of the past against the present. A short excerpt from the interview with Gencay, a woman from the trade notables, is illustrative:

> [Domestic chores] were a great burden. We had large lands. We had two horses for plough. We had also a man who used those horses and plowed the field. He took them in the morning and brought back in the evening. He tied them up. We did not see any of it. I mean they used to do it. We had a steward at the door and he did all. We never carried the burden. How should I tell? Those days were so good.[19]

Nostalgia is revealed in the notables' interviews as a yearning for a lost time. As Malpas argues, it typically implies an estrangement from the past, especially from childhood. It is as if the return to the past haunts individuals more often when the temporal distance increases—and typically, the nostalgic experience becomes more common among older people (Malpas, 2012, p. 167). Yet, these memories are also related to a sense of place. Events, people, and all sorts of details rebuilt by memories are inextricably bound up with the site of their happenings, from where they leap up and come to mind in a nostalgic experience. Their childhood reminiscences mark all these sites as places of memories. For example, the memories of Latife, about her father buying her the baby doll that she wanted so much, reveal days past in Aleppo:

> Now, we were in Aleppo and there was a shop called *Oeuvres de Pac*. Baby dolls that you could rock it to sleep or cry were all along the shop window. I used to sneak out and go to the shop, and I used to put my head on the window to stare at them. How childish! I could not think about asking to my parents to buy one for me. One day my father saw me there. "What are you doing here?", he asked. I was looking at the baby dolls. He said, "Shall I buy one for you?". I was surprised. "Come on, then! Which one do you want?" I liked the third one in the row. I do not know how much gold it cost. But he paid quite a lot and bought it.[20]

The home indicated in Latife's memories is an apartment flat in Aleppo where her family temporarily settled in 1919. As she mentioned, other local notables of Kilis, together with her family, had moved to Aleppo to stay during the tumultuous period of French occupation. Notables' recollections show that these families shared the same social world as the Aleppine notables. The male family members attended the same high schools, as there was no lyceum in Kilis town and had trade business with common associates. Notable women ordered fabric from the same shops and accepted invitations from Aleppine families:

> We used to go a lot [to Aleppo]. For example, I was six or seven years old. I had little siblings. We lived in Aleppo for a year and a half when the French occupied Kilis. [My father] had rented an apartment flat in the best locality. We had our farm in Syria after all. Everything we had came from there. Anyway, we went there [to Aleppo]. ... Now my parents were in contact with important families of Aleppo. There were a prominent one, called Hasip Effendi's family. They would go to his reception.[21]

In some cases, distinguishing between notable families in Kilis upon patrilineal origin might be a bias obscuring the roots of the family in both places. Several families in Kilis had established kinship relations through marriage with Aleppine families, as in the case of Latife's grandmother who came to Kilis following her marriage. The male members of younger cohorts could also endure the practice of taking a woman from Aleppine notables as an eligible spouse, when their wives had died. Or else, the kinship relationships could be sustained through polygynist practices of male members of notable families. Ayhan, a notable woman born in 1924, has four aunts in Aleppo since her grandfather had espoused his fifth wife, Ayhan's grandmother, in Aleppo. When she was a child, Ayhan was told that his grandfather used to sleep overnight alternately in Kilis and in Aleppo.

In these memories, the notables deploy a different image and meaning of their homeplace than those dominating in the present. Being notable for instance is implicated not only in the lifestyles and tastes that were associated with being civilized, modern, and cultured, but also in living in such a place. For Latife, "Kilis used to be a civilized place." Notables considered their homeplace as more modern in the past while the city of Aleppo, fully integrated with world capitalism in the early twentieth century, was for them the symbol of modernity. The words from Duran, though almost a Freudian slip, would support my assertion. As he explains when there was no border between Kilis and Aleppo, he says that "Aleppo was a province of Kilis, a province of the Ottoman Empire." Indeed, Kilis was then annexed with the province of Ottoman Aleppo. The notables' bonds extending across the border, strengthened by their proprietorship of landed estates, trade, and kinship relations are likely to undermine the imagination of their homeplace as curbed by a geographical boundary.

Concluding remarks

The chapters in this part have shown that the demarcation of the border may be both beneficial and disadvantageous to local notables. The closing of the Passavant gates and the confiscation of landed estates by the Syrian Baath Party in rule constituted a major blow to the local notables whose properties were left in Syria. With the disintegration of large landholding in Kilis countryside, the traditional landed notables are obliged to pay the cost of reproducing their class position by embracing what was once disgraceful for them: by circumventing the law. Trade notables more easily accommodated because of intergenerational occupational transmission. The other options for local notables to try and preserve their social standing were conversion of economic capital to social capital, i.e. education, out-migration, and urban employment, as well as marriage. Most notable families left the town and settled in metropolises.

Memories of notable families in this chapter have pointed to a nostalgic image of their hometown, which refers to the recollections about their life in their homestead in rural Aleppo or their visits and participation in the social life of Aleppo city. They have underlined that the meaning of living in Kilis for local notables signified a modern, urban, and cultured place because they shared a common world with Aleppo. These memories attested to the fact that local notables are the repositories of urban history in the provinces and their feeling of falling from grace is inevitably reflected in the urban decay, as they no longer have the determination or power to protect the historic heritage.

It is also argued that the political division between traditional landed and trade notables was accompanied by a symbolic struggle to maintain status distinctions.

While moral qualities ascribed to the members of traditional landed notables were influential markers of social distinction, trade notables could compete better in the field of culture and consumption. Notable families espoused distinctive life-styles and tastes considered modern and entered into competition with each other, where women played a leading role in their transmission. The following chapters show that the ascendancy of new wealth topples the social position of traditional landed notables, which the latter tried to secure against the trade notables. While the traditional landed and trade notables were not stigmatized as lawbreakers, the new wealth needs to restore its self-image to that of philanthropist businessmen, imitating old wealth in generating social capital with charitable activities in the town and, thus, naturalizing their dramatic ascendancy.

Notes

1 I observed that the old wealth was not likely to sell its ancestral patrimony, which explains why the property transfer concerned a building abandoned by the Armenians. The 1915 correspondence of the Directorate of Tribal and Immigrant Settlement with the Ministry of Internal Affairs gives clues about the abandoned properties confiscated in Kilis. For more detail see Çetinoğlu (2009).

2 This material was obtained by mixing the domestic solid waste burned in the bath furnaces (*külhan zibili*) with slaked lime and water and used for surface coating as well. Bebekoğlu and Tektuna indicate that artisans used to make colored patterns on this flooring that was washable and corrosion-resistant (Bebekoğlu and Tektuna, 2008, p. 170).

3 Yüksel, an old-generation woman married to a man from the trade notables in about her mid-seventies, has also shown her furniture bearing the signature of fine carpentry and woodworking as she kept her closet and mirror, as well as other furnishings.

4 *Bursa'ya gelin geldim. Şimdi Kilis deyince köy zannediyorlar ve benim bütün çeyizim Halep'te dikildi geldi. Çok şık şeyler.*

5 Can (2013); Demir (2006).

6 Also known as women's park, it was built in the late 1930s by the district governor clearing the ruins of the old mosque and cemetery at the town's center and seems to have served such a function. Interview with Gencay, born in 1929.

7 *Bu Ayşecik Parkı önceden çay kahve yoktu. Yalnız gidip oturulurdu. Onların, eşrafın yeri ayrı, normal ailelerin yeri ayrı. Yani kendilerini böyle şey etmişlerdi artık. Onlar kendiliklerinden yapıyordu bu işi. Onlara ayrı bir değer veren yok. Onlar kendilerini yüksek görüyorlar. Yani ben bunlara layık değilim. Ben bunlarla oturamam gibisinden kendilerine layık olan yere gidip oturuyor. Yoksa sen çok iyi bir ailesin, sen şuraya otur diyen yok. Küçük yaşta kendilerini öyle alıştırmışlar.*

8 The full text "Convention of Friendship and Good Neighbourly Relations Between France and Turkey" signed in 1926, also known as the Jouvenel-Aras Agreement, can be found in the League of Nations Treaty Series available at the World Legal Information Institute webpage: www.worldlii.org/int/other/treaties/LNTSer/1926/242.html.

9 Parliamentary question by Talat Köseoğlu (TBMM, 1966, p. 375).

10 Ibid, p. 375.

11 *İşte beni eletirler [iletirler] Halep'e. Babamlar giderlerdi. Ben de küçüğüm işte. 7-8 yaşındayım. Beni severler memurlar. İnerler yoklarlar. Babam da giy derdi. Bana bir önlük giydirirler. Önlüğün altına sade [sadece] altın dizerdi babam. Sade dizerdi. Beni aramazlardı, halbuki göyneğin altında altın var. Memurlar ha derlerdi, beni kucaklarına alır, öper öper bırakırlardı. Karışmazlardı. Halbuki bir şeyi geçirmezler ya amma. Beni severler deyi, benim arkama dizerdi rahmatlık. Küçüğüm, bak aklıma gelir.*

12   The proceedings about the speech of the foreign affairs minister Fuad Köprülü confirm that the Syrian government started to take a tithe from Turkish landholders in the early 1950s in reciprocation of land taxes taken from Syrian citizens by the Turkish government (TBMM, 1952, p. 123).

13   *Bizim de bir Hamparsum diye Suriye'de olduğu için zeytinlerimiz, o bakar alır gelirdi. Ondan haber geldi o gün. Kapıyı açıverdim, işte İş Bankasının üstünde oturuyorum. Gel kızım gel dedi. İndim aşağıya. Eskiden de çok serbest, tanış konuşulmaz ya. Buyur efendi dedim. Üvey oğlunu yollasan dedi. Yarın zeytini, alıcı bulmuş, yani toplamak için. Paranızı vereyim. Alsın gelsin. ... Şimdi o gün o şekilde kaldı kızım. Ben de sevindim. O zamanki parayla bize 6000 lira para verdiler. 6000 lira büyük bir paraydı 40 sene evvel. Aman sevindimki çocuklar da küçük, elimizde büyük bir para olacak. Akşam oldu. Kayınvalidem de var, görümcem de. Bizdeler. Hepimizin annemgile oturmaya gittik. Gece oturmasına gittik. Bu Raika da ordaydı. Kahve pişirdi getirdi bize. Oturuyoruk. Hiç unutamıyorum o günü yani. Daha yeni önceki gece uykum kaçtı aklımda o. Ondan sonra elimde böyle, çok iştahla içiyorum böyle kahveyi. O sırada kardeşim dışardan geldi. Yukarıya çıktı. Herkesin kahve fincanı okunmuş. Benimki elimde daha. Duydunuz mu dedi? Ne dediler? Hepimizin malı çünkü orda. Bugün Suriye'de hükümet el koydu, bitti dedi. Yani o gün. Birdenbire kızım. Yani ben o günü de unutamam. Birden eyvah öldüm dedim. Aha elimi şöyle etmişim. Acıdan nasıl vurmuşsam ki. Şu on parmağım ertesi günü simsiyah. O kadar mal gitti, o para, o 6000 lira da gitti. Onu da alamadık. Hayat öyle geçti gitti.*

14   *Zaten böyle bir çalışarak kazanma değil de hazırda oturup yeme alışkanlığı varken ve bu da miras yoluyla parçalanınca ve bu beyler de kaçakçılık da yapamadıkları için çünkü onlara göre ayıptı. Bir naylon çorap giydirmedi babam bize. Suriye'den o zaman naylon çorap gelirdi, biz giyemezdik. Bizim köy tam hudutta olduğu halde, bizim arazinin olduğu yer kendi elemanımız, vekilharcımız, yarıcımız kaçakçılık yaparken biz diyemezdik ki kardeşim şu yüz lirayı da bizim için al, bizim de hissemiz gitsin gelsin. Demedik, daha doğrusu utandık. Dolayısıyla onlar kazandı. Bire on, bire yüz. Biz yüze bir, yüze beş kazanınca tevercinin kefesi dağıldı. Sonra okuyan tekrar Kilis'e dönmedi. Doktor oldu, İstanbul'da kaldı. Avukat oldu, İstanbul'da kaldı. Ticaret yaparken Adana'ya taşındı.*

15   Though the interviewees emphasize the metropolitan cities as a migration destination, it is worth noting that Gaziantep city has the biggest migrant population, probably with a lesser yet remarkable proportion of notable families. See, the Turkish Institute of Statistics Address-Based Population Registration System database for the destination cities with the highest rate of Kilis migrants for 2011, respectively: Gaziantep 89,918; Istanbul 35,934; Ankara 5,307; Adana 5,164; Mersin 4,097; Antalya 3,394; İzmir 3,199; Bursa 1,587.

16   For instance, the interview with Cemil, a former senator, dwelled mostly upon his ideas of promoting the town's development through cultural, religious, and health tourism by taking advantage of the town's assets in terms of historic heritage and Mediterranean microclimate.

17   [Kaçakçılığın başlaması] *mayınlı saha döşendikten sonra 1954-56'lara rastlar. Zaten ondan evvel de aramızda sınır yoktu bizim. Kilis'le Halep arasında sınır yoktu.*

18   *Hayır o zaman yoktu. Zaten Halep Kilis'in, Osmanlının vilayetiydi. Mal değiş tokuşu, ticaret yapılırdı. Oradan mesela ipek kumaş gelir; sine satin, enver satin isimleri çoktur. Hudut yoktu. Buradan Antep'e gidip satar gibi Halep'ten getirir Kilis'te satarlar, Kilis'ten götürür Halep'te satarlardı.*

19   *Çok ağır işti. Bizim çok arazimiz vardı. Evde ayrıca çift denilen iki tane at vardı. Bir de o atları kullanan, tarlayı sürmeye giden adam vardı. Sabahleyin gelir onları alır götürür, akşam üzeri geri getirir. Yerine bağlar. (Biz) onların hiçbirini görmedik. Yani adamlar yapar, kapıda bir uşağımız var yapar. Biz o sıkıntılara girmedik. Ne bileyim nasıl anlatsam, çok güzel günlerdi.*

20 *Şimdi Halep'teyiz, orada Ovr dö pak diye bir mağaza var. Vitrininde boydan boya bebekler var, şu uyuyup şey eden, ağlayan. Evden kaçarım, gider vitrine başımı koyup o vitrine bakarım. Çocukluk yani, al demek aklıma bile gelmez. Bir gün babam gördü beni orada. Ne yapıyorsun dedi. Bebeklere bakıyorum. Sana alayım mı dedi. Şaşırmışım. "Gel öyleyse içeri" dedi, "hangisini istersin?". Baştan beri üçüncü bebeği çok beğenirdim. Valla kaç altındı bilmem. Epey bir para ödeyerek aldı.*

21 *(Halep'e) Çok giderdik biz. Mesela ben altı, yedi yaşındayım. Bir tanesi 6 aylık, bir tanesi kırklık. Kilis'e Fransızların geldiği zaman biz Halep'te bir buçuk sene oturduk. [Babam] en güzel yerinde bir apartman dairesini tutmuş bize. Zaten çiftliğimiz Suriye'de. Her şeyimiz ordan geliyor. Oraya gittik ... Şimdi Halep'in büyük aileleriyle konuşuluyor, Hasip Efendiler diye meşhur bir aile var. Onun gününe gidecekler.*

References

Aswad, B. C. (1974) "Visiting Patterns among Women of the Elite in a Small Turkish City", *Anthropological Quarterly: Visiting Patterns and Social Dynamics in Eastern Mediterranean Communities*, 47(1): 9–27.

BCA (1931) Başbakanlık Cumhuriyet Arşivi (The State Archives of the Prime Minister's Office), 180/244/6, 5.12.1931.

Bebekoğlu, S. and Tektuna, M. (2008) *Kilis kültür envanteri: kentsel mimari, kırsal mimari, yazıt eserleri*, Kilis Valiliği.

Bertaux-Wiame, I. and Thompson, P. (1997) "The Familial Meaning of Housing in Social Rootedness and Mobility: Britain and France", pp. 124–182 in I. Bertaux-Wiame and P. Thompson (eds.), *Pathways to Social Class: A Qualitative Approach to Social Mobility*, Oxford: Clarendon Press.

Bouchair, N. (1986) *The Merchant and Moneylending Class of Syria Under the French Mandate, 1920–1946*, unpublished PhD thesis, Georgetown University.

Can, E. (2013) "Shifting boundaries and dynamics of neighbourhood: Women in public parks in Ankara", paper presented at International Resourceful Cities 21 Conference, 29–31 August.

Çetinoğlu, S. (2009) "Ermeni Emval-i Metrukeleri Üzerine", *Birikim*, June 8.

Demir, E. (2006) "Toplumsal Değişme Süreci İçinde Gençlik Parkı: Sosyolojik Bir Değerlendirme", *Şehir Plancıları Odası Planlama Dergisi*, 4: 69–78.

Karadağ, M. (2005) *Class, Gender and Reproduction: Exploration of Change in a Turkish City*, unpublished PhD thesis, University of Essex.

Karadağ, M. (2009) "On Cultural Capital and Taste: Cultural Field in a Turkish City in Historical Perspective", *European Societies*, 11(4): 531–551.

Kesici, Ökkeş (1994) *Kilis Yöresinin Coğrafyası*, Ankara: Kilis Kültür Derneği Yayınları No. 12.

Keyder, Ç. (1991) "Introduction: Large-Scale Commercial Agriculture in the Ottoman Empire?", pp. 1–16 in Ç. Keyder and F. Tabak (eds.), *Landholding and Commercial Agriculture in the Middle East*, Albany, NY: State University of New York.

Malpas, J. (2012) *Heidegger and the Thinking of Place: Explorations in the Topology of Being*, MIT Press.

Morley, D. (2001) "Belongings: Place, Space and Identity in a Mediated World", *European Journal of Studies*, 4(4): 425–448.

Newman, K. S. (1999) *Falling from Grace: Downward Mobility in the Age of Affluence*, Berkeley: University of California Press.

TBMM (1952) TBMM Tutanakları (Proceedings of TGNA), Vol. 17 Session 4, 14.11.1952.

TBMM (1966) TBMM Tutanakları (Proceedings of TGNA), Vol. 2, Session: 34, 12.1.1966.
TBMM (1968a) TBMM Tutanakları (Proceedings of TGNA), Vol. 28, Session: 74, p. 17, 10.6.1968.
TBMM (1968b) TBMM Cumhuriyet Senatosu Tutanakları (Proceedings of TGNA Republican Senate), Vol. 45 Session: 31, p. 1012, 8.2.1968.
Wilson, T. and Donnan, H. (1998) *Border Identities: Nation and State at International Frontiers*, Cambridge: Cambridge University Press.
Yeoh B. and Kong, L. (2006) "The Notion of Place in The Construction of History, Nostalgia and Heritage in Singapore", *Singapore Journal of Tropical Geography*, 17(1): 52–65.

Part II

New wealth and the middle class

4 Wealth generation and the rise of new rich in the margins of economy and state

This chapter focuses on the growth of the shadow economy along Kilis border and the rise of new wealth since the 1960s. It first introduces the case of a former smuggler from Kilis whose 'success story' of becoming a decent wealthy businessman. The new wealth of Kilis becomes visible in the post-1980 period, especially by benefiting from the state incentives promoted by the Özal government to subsidize export-led growth. I develop my discussion by introducing a theoretical framework on the transnational shadow economy in order to demonstrate from the vantage point of the businessman's story that illegal trade enables upward mobility and secure welfare unequally among socio-economic strata in the town.

Then, I shift into the 1960–1980 period of import-substituting industrialization (ISI) in order to trace the roots of wealth generation in Kilis. My analysis problematizes what the political scientist Peter Andreas calls "illicit globalization" (Andreas, 2011; cf. Tagliacozzo, 2007). Andreas argues against the strong conviction among globalization scholars that the nation-states lose their control over illegal transnational business transactions in an increasingly globalizing economy. A historical approach, however, reveals that neither the dimensions nor the magnitude of illicit globalization can be credibly demonstrated to have been augmented since the last century and "transnational organized crime is simply a new term for an old economic practice"—i.e. smuggling (Andreas, 2011, p. 406). Despite their diversity, these transactions actually have the same features of circumventing law enforcement by using unregulated and unauthorized channels of circulation and moving across borders in ways to elude detection and comprehension. In this regard, the growth of the transnational shadow economy foreshadowed the neoliberal trade liberalization in Turkey.

This chapter aims to offer a reading of the ISI period from the perspective of Kilis border. The characteristics of this period were the management of the economy by the state in order to protect the domestic manufacturing bourgeoisie from international competition and secure a redistribution of income to constitute a domestic market (Keyder, 1987, pp. 150–151). Thus, ISI referred to a social welfare regime with the central role of state in allocating scarce resources. This period was characterized by the imposition of strict regulation on the circulation of foreign exchange as well as high import tariffs and quotas in order to provide favorable conditions for domestic industrialization.

Diverging away from approaching the development of informal/illegal markets as a failure of state policies, this chapter demonstrates how dwellers are incorporated into the redistributive logic of the shadow economy. Border dwellers took benefit of the protective measures by reckoning rents to the illegal entry of consumer goods as well as gold and foreign currency and made their living in a border region that lacked opportunities for regular employment and a secure salary within the framework of ISI-based development. I support these points by focusing this time on the story of an extended middle-class family, influential in local politics, which grew rich from a poor rural background to own a transport company and other enterprises. From the vantage point of a family story, I show that the illegal trade of gold as well as consumer goods initiated a pattern of vertical mobility that moved rural families to the ranks of middle urban strata and overcame the structural constrains of social stratification in the town.

Story of a barber: smuggler, crime boss and businessman

> No one can call me heroin dealer and arms dealer. I grew out of barbering in the same way as Vehbi Koç grew rich out of grocery and Sabancı grew rich out of porterage.[1]

These words came out in a published interview from a Kilis native a decade ago, an alleged crime boss notoriously known for dealing gold, heroin, and arms with connections to the international underworld and later recognized as a decent businessman after he was cleared of all corruption charges with minor criminal fines. The reference to Koç and Sabancı, the two richest capitalists in Turkey, is meaningful. These two figures are suspected to have grown rich thanks to state patronage, governmental incentives, and favors during the first half of the Republican Era, while there is a strong public conviction that they are exemplary figures, hard-working and astute businessmen, who earned their wealth through their own efforts. Their careers were success stories of men who rose up from the lowest occupations—being a grocer or a porter—to the top positions in business. As such, their stories are supposed to be a moral for ambitious individuals who dream to strike it rich, teaching them to work hard so that they will be deserving. But they also promised the possibility of becoming rich to everyone who aspired and committed to work for it.

The barber, who assumed this nickname, had moved to İstanbul where he continued his investments in the lodging industry after he was cleared of the charges he was prosecuted for by the military government in the 1980 coup. Now heading Kilis Vakfı (Kilis Foundation)[2] in İstanbul, he is well known in his native town for his philanthropic activities conducted via the foundation, such as sponsoring the education and boarding of native students in the cities, and poor relief and donations for mosque repairs, public school and university buildings. He makes frequent appearances in his hometown and he is held in high esteem by the local authorities.

The barber is the most popular name among the benevolent businessmen who were based in İstanbul but known for their charity work in their native town (See Figure 4.1). His wealth is not inherited nor enabled by occupational transmission from his father. The nickname barber remained with him from his adolescence when he worked as an apprentice to his father's barbershop. It is suggestive of his success in climbing the social ladder from bottom up in the span of a lifetime, after the early death of his father. In addition, it was the obvious indicator of his past, where he had come from, arousing curiosity as to how he could make such a fortune by beginning his business career as a barber's apprentice. He succeeded in making his wealth, unsanctioned according to law enforcement bodies during the decade of the 1980s, which was accepted as normal and established his social position as a decent businessman.

This transformation in image was also relevant for the broader new wealth of the town that, since the 1960s, prospered from smuggling. As a former mayor and parliamentarian of Kilis indicated, the brigadier general of Martial Law Command in Gaziantep gave the town dwellers suspected of smuggling a hard time and picked up almost everyone, leading eventually to the outmigration of the wealthier. A former smuggler interviewee complained that the military government raided the new wealth and pictured them as *baba*, meaning both father and crime boss in Turkish. When the newspapers resurrected the serious allegations

Figure 4.1 A modern minaret added by the barber as a philanthropic activity, 1978
(Photo courtesy of author)

about transnational underworld relations reaching out the then Prime Minister in late 1980s, Kilis would be mentioned as "little Sicilia" because of the many Kilis natives who, among the "bosses", were involved and who made prominent appearances in the latest news coverage and had kinship and marriage relations to each other, including the barber (Milliyet, 1988).

The news covered the fact that the barber was confronted several times after the coup with charges of gold smuggling, money laundering, fraudulent export, and instigation to murder and, when faced with arrest warrants, stayed in Switzerland where he had a residence permit; he was never convicted of these charges (Yetkin, 1989). Upon one of his returns to Turkey, being pardoned by then Prime Minister Turgut Özal, the only penalty he would get was a fine of 100,000 Turkish Liras for violation of the law regarding the protection of the value of the Turkish currency thanks to the amendment—though the charges of gold smuggling initially threatened him with a possible prison sentence of 20 years.

The barber's name reappeared after the recent exposure of a police operation into bribery and corruption on December 17, 2013, which captured the headlines and put the current AKP government under pressure to declare it a dirty plot to terminate its administration. The allegations pointed at a money laundering and gold smuggling ring between Turkey and Iran, involving several sons of cabinet ministers, leading businesspeople close to the government and executives. Reza Zarrab, a young Iranian businessman and gold dealer, was implicated as a ringleader, exporting $6 billion worth of gold from Turkey by dodging economic sanctions against Iran over the last year and distributing $60 million to his Turkish partners for enabling the money transfer through the accounts of real or front companies in Turkey.

Following the scandalous leak of the visuals and telephone recordings about the anti-corruption probe, the barber's name was mentioned as the person who introduced the Iranian businessman to the Minister of Interior Affairs, who was forced to resign together with two other cabinet members a week later after the probe hit the headlines. Thus, a variety of columnists drew probable continuities and links of similarities between the anti-corruption probe and a series of investigations concerning money laundering and gold smuggling by a Turkish–Bulgarian ring in the late 1980s, in which the barber was alleged to have a more pivotal role.[3] The ring was implicated in smuggling nearly 450 tons of gold from Turkey to Switzerland, where a Lebanese–Armenian trader–moneylender was allegedly the ringleader of the Lebanese connection. The ring was also suspected of laundering illicit money for financing drug and arms trafficking, as the illicit money was transferred back in the form of gold and foreign exchange smuggled into Turkey via the Bulgarian border.[4]

It was argued that the gold smuggling in the first three years of the decade, benefiting from the price differentials between the two countries, was a huge detriment to the Turkish economy. A bulk of foreign currency transfer could also be shown as pre-financing payments for fraudulent export from Turkey, as an investigation into fraudulent export suggested $55 million worth of transactions had been made. The ring members also benefited from tax rebates in exchange for the currency transfer much needed supposedly by the Özal government in order to finance the country's development—the gold and foreign exchange market was newly liberalizing in late 1980s Turkey and their influx, arguably, was a necessity.[5]

A corruption scandal blew wide open with the news of Prime Minister Turgut Özal's meeting in Zurich with prominent Turkish and Lebanese underworld figures—about whom arrest warrants or red notices were issued by Turkish judicial authorities—and outstanding Turkish businessmen including the soon to-be chief executive officer (CEO) of a Turkish public bank.[6] The barber was alleged to have organized the meeting in order to bring the Lebanese–Armenian trader and the Turkish Prime Minister together. Later on, the barber would confirm the meeting without revealing its context and asserted that the Prime Minister asked the Lebanese–Armenian gold dealer, who was the top trader in the world in those days according to the barber, to set up the wholesale gold market in Turkey, the proposal being refused by the gold dealer.[7]

The money laundering had broader repercussions in the Swiss press. The muckraked news indicated that the money laundered in Swiss banks by the Lebanese connection was over $2 billion. The corruption upheaval toppled the Minister of Justice, seemingly destined to be the next president of Switzerland, from her post, though ironically she was a supporter of the anti-corruption law foreseeing the transparency of banking accounts and several other measures. She had to resign in early 1989 after the allegations that she tipped off her husband about a potential money laundering investigation regarding the Lebanese trader's company, where her husband was the vice chairman. The Lebanese trader received no conviction in Switzerland, but he had to move to Dubai after the reforming of the Swiss banking system, ending the country's status as a safe haven for illicit money.

Wealth generation in the margins of economy and state

Our knowledge of the barber's story will always be limited with the tip of iceberg exposed in the Turkish media, knowledge which is merely surrounded by inconclusive police investigations and parliamentary inquiries, and ambitious journalistic accounts with bits and pieces from intelligence reports about 'the deep state.' Nevertheless, it is plausible to suggest on the basis of hearsay evidence I collected during the field study that the economic accumulation in Kilis allowed the growth of conditions for the new rich to move up the social ladder and to acquire unprecedented success in the eye of their townsmen. Thus from the vantage point of the barber's story, I aim to provide an insight into wealth generation in Kilis. My intention is not to prove the 'fraudulent ways' that the new wealth of Kilis have resorted to, but to provide a framework for exploring the possible trajectories of its social reproduction. The wealth generation in Kilis town was connected to the expansion of transnational shadow networks along the Turkish borders, which moved the local actors to a transnational scale after the introduction of post-1980 neoliberal policies. As illustrated by the barber's story, the new wealth deserted the town and Kilis lost its salience as a crossing point after the military coup.

The interviews with the town locals indicated that there was already trafficking of gold and foreign currency in place before the Özal years, traded legally to Beirut, then smuggled from to Istanbul via the Kilis border.[8] The gold money in return was clandestinely transferred via the same route. Beirut was renowned then as a major center in the international gold trade since the 1950s and the "Switzerland of the Middle East" thanks to the adoption of the Swiss-style Bank Secrecy Law in 1956 (Gates, 1989, p. 19). The national gold and foreign currency market, with its heart at Grand Bazaar in Istanbul, was largely underground in the pre-1980 period, as their purchase as well as trade was strictly regulated by state tariffs and impositions.[9]

The gold-jewelry shops in the Bazaar not only provided smooth dealings for gold trade, but also ran "a parallel banking catering system, on the demand side, mainly to businesses seeking illegal foreign exchange and on the supply side, to people wanting to quietly convert foreign exchange they were bringing to Turkey" (Naylor, 2004, p. 201). This banking catering system also served as a monetary deposit based on trust, whereby the international money transfers could be made clandestinely.[10] Thus, the system was feasible for smugglers aiming at transferring drug money, as well as other immigrant workers seeking a better exchange rate and evading having to pay tax for importing their remittances.

The allegations that the barber's relationship with the Lebanese trader dated back to the early 1960s seems to confirm the interviewees' claims about the illegal gold trade across Kilis border. The barber had business relations with his father in Beirut, whose international reputation as a gold dealer attracted smugglers from Turkey as well until the father and son moved to Switzerland in the mid-1970s with the outbreak of the Lebanese civil war. According to the hearsay evidence in the press, the barber was doing trade with senior and junior Lebanese traders and controlled the gold and foreign currency traffic from his office in the Grand Bazaar, where Kilis businessmen were amongst the major players of the market in the 1970s, as my interviewees asserted.

On the other hand, the barber's story illustrates the fact that the border regions like Kilis town are not marginal. As Roitman showed us in her study on the economic accumulation in Chad Basin (2004c), wealth generation on state margins depends on wider commercial and financial relations at the metropolitan centers. The barber's cohort are among the new class of entrepreneurs rising on the national horizon with the introduction of neoliberal economic policies by Özal after the military coup, which led to the substantial growth in the income gap between the rich and the poor in Turkey. The rise of new wealth is problematized by the scholarly and media discourses.

The scholarly discourse tends to regard the corruption as an anti-thesis of free market development, a disease associated with the mentality of state authorities and needed to be cured particularly in underdeveloped and developing countries. Öniş, for instance, assumes that the Özal legacy in the post-1980 era signified the feeble commitment to the rule of law and norms of a democratic polity and the irresoluteness for establishing the legal infrastructure for a well-functioning liberalized market economy (Öniş, 2004). The problem with this approach is its lack of comprehension that the intermingling of the state regulations and formal economy with the criminal networks and informal markets is characteristic of neoliberal governance (Roitman, 2004c; Nordstrom, 2000; Galemba, 2013). Just as the Turkish interim government after the military coup of 1980 expanded its fiscal base by subsiding export-led growth, while undermining the social and economic rights to the expense of widening the income gap, other neoliberal states of the "Third World" abetted the economic activities of criminal networks in order to finance their development.

On the other side, the corruption issue is mostly studied in Turkey by journalistic accounts, which gave extensive reference to the police and intelligence reports and indicated the relationships between state bureaucrats and 'mafia' as the causes of corruption (Hatip-Karasulu, 2005). These accounts adopted a more critical perspective towards the so-called free-market ideal. But, they assumed a notion of the state "as a unitary, preordained actor" (Galemba, 2008), even though they implied the pluralization of regulatory authority within the state, indicating the infiltration of the 'mafia' into state bodies. Carolyn Nordstrom (2000) defines these types of economic activities as a shadow economy. However, the shadow economy, as Nordstrom argues, is based on vast extra-state networks expanding across all the world's countries (Nordstrom, 2000, p. 36). This does not mean to say that these networks are not working through and around formal state institutions. But Nordstrom encourages us to conceive the shadow economy as more formalized, integrated, and bound by rules of conduct than we tend to think.

She also objects to the distinction between state and non-state, formal and non-formal power relations, reminding that in markets and people's lives what we regard as criminal and illegal activities are closely integrated with mundane efforts of earning livelihoods (Nordstrom, 2000, p. 40). To adapt Nordstrom's argument to the case of gold smuggling through Kilis border, it would not be true to discuss it as a distinct sphere of activity, separate from the lives of peasants who carried it with them by sneaking across the mined zone, or the relatives of local bosses who are entrusted to keep the smuggled goods in deposit until they are shipped to the next destination, or the drivers who deliver these goods to the bigger bosses in the cities.

So far I have introduced a framework for discussing how wealth generation in the margins is encouraged and facilitated by the reconfiguration of governmental relations in the post-1980 era. The following section will explore the growth of the shadow economy during the import substitution period and demonstrates how the nascent markets of Kilis town are articulated with the transnational flows of goods and money. I will focus in the following section exclusively on the story of an extended middle-class family influential in local politics, who grew rich from poor rural background to own a transport company and other enterprises. The family's story will allow me to trace the history of gold smuggling as well as other contraband to the 1960s. Drawing on the family's story, I will demonstrate that the illegal gold trade initiated a pattern of vertical mobility that moved rural families to the ranks of the middle urban strata and overcame the structural constraints of social stratification in the town.

Upward mobility at the intersection of small-scale trade and transnational crime

The Koyuncu family owned a large apartment building in the town center. The ground floor of the apartment building was used as the office of the transport company for national and international shipping. A door at the rear connected the spacious office place floored with old furniture to a very large kitchen, turned into an almost charitable soup kitchen every Friday because of the usual regulars showing up after the Friday prayer at the mosque and occasional visitors in need of hot soup. The whole building accommodated the households of several brothers from the Koyuncu family. So, the extended family practically lived together. The proximity of the apartment building to the largest mosque in the town made the company office a popular place for socializing, as Rıdvan, the buffet owner across the three-way intersection within full sight of both the apartment building and the mosque, told me. I had a habit of dropping by the shop on my way to the town center from time to time for the sake of tea offered by the owner and small talk. Though a humble and meek man, he used to lose his temper when he started talking about the country's agenda. These were the times when the conversation led to the Koyuncu family.

The buffet owner could not stand the sight of town dwellers crowding into the company office to ingratiate themselves with the family members. He was annoyed by the fact that several families in the town prospered from the shadow economy with the help of political patronage relations. The Koyuncu family, who went from a peasant origin to rich, local entrepreneurs, was an influential one in local politics. Some of its members held eminent positions in the local organization of AKP, the party in power, although the family was known as the loyal adherent of a former center-right party until the death of the head of the family, the elder of several brothers and the chairman of the local party branch. A resident of the border village, where the Koyuncu family had its origins, remembers vividly that when the family head hosted the village party leader in the mid-1990s, who happened to head the cabinet as well, the women of the village became sick because of laboriously cooking food and honouring their important guests.

İsmail, an elder brother of the Koyuncu siblings who I met in the company office, was a man who wore traditional clothes and baggy trousers, reminiscent of the social background of the family despite its upward mobility. The family, as previously mentioned, had its origin in a border village where Arabic and nomadic Turcoman tribes with Kurdish relatives due to ethnically mixed marriages were settled.[11] The demarcation of the Turkish–Syrian frontier turned these populations into extended families with kinship relations straddling the border. The elder brother indicated the family was a stockbreeding agha family, though it became impoverished through the loss of lands and cattle after World War II. He did not abstain from saying that the family was involved in trading contraband gold before the military coup of 1980. The gold trade across the border is known to date back to 1940. The gold trade remained regional at the beginning: it circulated from Beirut to Aleppo, where the local gold dealers sold it to buyers from Turkey.

The family business began with their father and consisted of taking delivery of contraband gold at the border and carrying it to the city of Gaziantep. These were 31-ounce (nearly a kilogram) gold bullions officially traded from Switzerland to Beirut and clandestinely trafficked to the Syrian inlands and across the Turkish borders. In the 1970s, the family started to deliver the gold to Istanbul Grand Bazaar, where entrepreneurs from Kilis established their business and the gold market expanded to a national level.

But I gathered the family's story particularly by interviewing another member who moved away from his family, being offended by a kinship discord some years ago. Mustafa, repairing clocks in an arcade store, differed from his brothers in terms of political view and outlook. The interview offered a broader perspective to explore the growth of a semi-legal market that brought major changes to the social stratification in the town and its urban landscape, as he probably felt less attachment to his origins and more independence to speak out. Mustafa remembers that he was reckless in his early adolescent youth in shuttling across the border to smuggle jackets left over from Levantine citizens and the earlier presence of Western soldiers. The village was a small one with 14 households. The interviewees' recollections indicated the seesaw movement of villagers across the border before the mining of the frontier zone. Mustafa remembers that as an adolescent he received 2.5 to 5 Liras for each delivery of gold crossing the border—when a kilogram of gold cost 5,000 Liras. Three kilometers between the border villages at different sides of the frontier and kinship relations made frequent deliveries possible. He picked up the delivery left at the bottom of a border stone, used to mark the frontier line until the late 1950s, and drove it to the local jewelry dealer on an old Russian husky motorcycle, perfectly steering along the village roads. The local jewelry dealer would convey it to its next destination. The local gold dealers kept account and Mustafa received his accumulated earnings once every few months.

Smuggling across the border was a dangerous business since the illegal crossings could lead to armed clashes between soldiers patrolling the border and the villagers. Yet, the danger threat peaked with the planting of land mines, turning a width of 400 to 800 meters along the frontier line into a field of death. The price of gold per kg as well as the earnings of porters, called sırtçı (literally translated as piggybackers since the hired porters carried the load on their back), went higher and higher. Moreover, the volume of gold trafficking significantly increased, with the great majority of bullion smuggled into Turkey crossing Kilis border and a slight percentage circulating across other points on the Turkish–Syrian frontier. The price of a kilogram doubled in the 1960s and saw a steady increase in the 1970s.[12] A porter could earn 100 Liras for the load he carried—usually it did not matter what the type of load was. Mustafa's brothers assumed the position of local gold dealers and their share in the gains mounted to 600 to 700 Liras per kilogram, minus the bribe given to the border patrolling soldiers. For Mustafa, this meant assuming the position of a boss administering the gold trade in contrast to the status of a porter whose labor force was employed by someone else, though the Syrian trader and final Turkish buyer actually arranged the details of commerce.

In the late 1970s, about 300 to 400 kg of gold crossed the border daily, often by porters who shuttled on wheel back and forth across the border gate rather than sneaking into the fenced minefield.

The story of the Koyuncu family illustrates the rapid upward mobility of a lineage from a rural background into the ranks of the urban middle class by appropriating kinship and political patronage as an asset and developing the father's occupation of smuggling into a family business. The family filled a significant position in local politics, entangled in an urban economics of reconstruction and clientelism of social assistance. During the fieldwork, I observed that the political clientelism was effectual in the distribution of rents tied to the urban development as well as of social assistance benefits. The family's move into the ranks of the middle class needs to be fitted within the broader picture of gold traffic from Switzerland to Istanbul via Beirut and Kilis, complemented with the flow of foreign currency—the gold money—that reached Switzerland through the same route. Enver, a former gold smuggler and fellow villager of the Koyuncu family, expressed in his interview not only the role of Kilis townsmen in the Grand Bazaar, but he also gave a glimpse into the concomitance of the gold and foreign currency trade:

> We used to scale the Bazaar up and down. Once, I saved my money on the safe box of a fellow townsman until the goods that I had bought on loan would be sold... He came a bit late. I filled the box with German marks and American dollars until he came. I collected a wad of currency in the Grand Bazaar of Istanbul.
>
> ...
>
> You take away gold from here and exchange it with foreign currency there.[13]

The smuggled gold was sold in Istanbul in exchange for foreign currency and the latter clandestinely circulated to Switzerland by taking it out of Kilis border.[14] Except for the yearly 500 to 600 kg of gold production, the overwhelming majority of the national gold market was fed by contraband trade during the 1960–1980 period and only in the last years of the period lower-quality Iranian coins and gold bars entered along the borders of the country (Sağlam, 1991, p. 65). This made Kilis natives major national dealers in the gold trade, closely intertwined with the underground foreign exchange market and, among them, few could even wriggle out as crime bosses running a business with their transnational counterparts.

As the stories of upward mobility demonstrate, the gold trade is a useful case to reveal the subtle and hazy connections between transnational crime organizations and small-scale cross-border trade. As Yükseker discussed within the context of informal trade between Turkey and Russia, transnational informal trade is open to concentration of capital and monopolization (Yükseker, 2003, p. 67–68). Thus, they can lead to the emergence of crime organizations, which are able to acquire monopoly or semi-monopoly rent and take advantage of patronage relations with the state and use of violence, operating in high-risk areas where they have few competitors. Within this larger context, the subjective perceptions of the Koyuncu family testify to the tendencies toward the concentration of capital and monopolization even more dramatically. As profit rose from increasing business and the involvement of long-distance legal and illegal buyers, border dwellers in such lucrative businesses were less likely to abide by redistributive norms and created family monopolies around their respective enterprises (Galemba, 2012a, p. 10). The family grew their gold business from regional small-scale trade up to the underground domestic market. Mustafa's words hinted at how he and his brother benefited from the local structures of opportunities and established themselves as local entrepreneurs, exploiting the labor of their fellow villagers: "We got promoted and we moved the 'company' from the village here [to the town]".[15]

The Koyuncu family illustrates that the gold trade was significant in the moving of rural families into the ranks of the middle urban strata, which grew richer under the auspices of new wealth. So, the social mobility of the family should be understood as part of the shift in social stratification structure that gave rise to the new wealth, gradually undermining the economic and symbolic status as well as political dominance of the old wealth. The enrichment through gold trade marked the involvement of rural families among the middle urban strata, as a member of these families could acquire a nickname denoting richness after they achieved this coveted status.

As Sarah Green argues, gold signifies for border dwellers in a state of flux, as well as for broader Eurasian people, the most obvious embodiment of richness and a stable means of preserving the family wealth (Green, 2009). For instance, a former help from a rural background who worked for an enriched gold trader, succeeded in yielding enough fortune to own a stone courtyard house, which is still named after him.[16] This example illustrates the emergence of new wealth that could move up the social ladder bypassing the regular patterns of mobility such as intergenerational occupational transmission or access to education.

The following section will extend the analysis of these mobility patterns to the town landscape and domestic market, by focusing on the decline of traditional industry, urban development and the scaling up of the local economy to the national level. I will depict how the unregulated flows of goods transformed the town landscape into a border zone speckled with high-walled warehouses of transport companies, small shops, arcade of stores, and houses used as bulking and diffusing points. Besides, I will underline how the shadow economy of 1960–1980 forged a local semi-legal market, bringing informal and illegal activities together, dependent upon commercial and financial relationships with the cities.

Concluding remarks

The chapter contemplates the way in which wealth is generated outside the formal state and market channels in the margins of economy and state. Against the background of ISI-based development policies, it explores the economic and political mechanisms by which the economic accumulation through illegal means is rendered as socially accepted and rightfully earned wealth in Kilis town. It shows that the contestation of the geographical border does not put state sovereignty and the domestic market at stake. As the story of the barber illustrates, local trade markets interweave into transnational shadow networks, which expanded across the trade and finance centers of the world in the 1960s.

As long as the transnational shadow economy in Kilis develops into a distinctive redistributive mechanism, wealth accumulation fosters success stories rather than incrimination of new rich, in contrast to the landed notables; a factor that compels the latter to mimic the social mobility strategies of the former and negotiate new alliances, i.e. inter-marriage. The chapter also makes clear that illegal trade activities tend to extend to the broader local community, involving notable families, large-scale entrepreneurs, extended families with a rural background, artisans, and border villagers. The cross-border trade fills the lack of investments and smoothes the workings of uneven capitalist development by providing employment in the town.

Notes

1 "*Bana kimse eroinci, silahçı diyemez. Vehbi Koç bakkalıktan, Sabancı hamallıktan zengin olduysa, ben berberlikten gelmişim*" (Mercan, 2002).
2 A benevolent association of Kilis migrants established in 1993 mostly recruiting among the new wealth that had to desert the town after the 1980 coup. I have been told by the town dwellers that the military government hampered the legitimacy of their smuggling business by criminalizing them and obliged them to leave the town to re-settle their business in Istanbul. The efforts to establish their reputation as absentee notables through philanthropy seemed to be successful.
3 See Dündar (2013); Erdoğdu (2013); Kahraman (2013); Hiçyılmaz (2013).
4 Sağlam (1991, p. 63) argues that the Bulgarian government facilitated the illegal trafficking of gold and foreign currency across its border by establishing a company to take a commission of USD 50 for a kg of gold or 1% for currency transactions.
5 According to Webb and Öniş tax rebates worked as subsidies because "first, the subsidy rate was not related to the total amount of taxes paid by the exporter and could exceed it. Second, the rebate scheme was introduced before the value added tax; when the actual value added tax rebate was added, the prior rebate scheme remained as a pure subsidy" (Onis and Webb, 1994, p. 157).
6 The news resurfaced in 1989 but the date of the meeting is not clear. The article series published in *Milliyet* did not state the date of the meeting, but gave place to the statements of the Lebanese trader confirming the meeting. An article by Erbil Tuşalp in *Birgün* declared the meeting was in the summer of 1985 (Tuşalp, 2005).
7 In fact, the allegations went far beyond and it was argued that the Prime Minister also proposed dual citizenship to him and an incentive to let him open a bank in Cyprus, while the gold dealer declined the proposals of his proponent. The barber claimed that the Lebanese–Armenian trader did not have any illegal business with Turkey and he sold gold to the Turkish Central Bank during the Özal years. Though the governments attempted to liberalize the gold market in 1984, the Central Bank owned the monopoly status until 1989 (Sağlam, 1991).
8 This is also supported by research on the gold trade in Turkey. Drawing on the interviews with exchangers from a Kilis origin in the Grand Bazaar, Sağlam points to the leading role of Kilis traders in facilitating national and transnational gold traffic. But in the early 1980s, the transnational gold trade mainly consisted of collecting the low-purity Iranian gold in the domestic market and processing it to gold of a standard value. Still, the traffic of gold coming across Syria continued until the Syrian

government restricted the entry of precious metals along its borders in the mid-1980s (Sağlam, 1991, p. 63).

9 It is not possible to estimate the volume of gold sold out in the Bazaar before trade liberalization, but the underground sales of the Bazaar reached about 200 to 250 tonnes between 1980 and 1982 (Sağlam, 1991).

10 I presume that the unregistered exchange office-jewelry shops in the town functioned similarly, by selling foreign exchange at the market rate and facilitating the money transfers outside the confines of state regulations. As in a recent anecdote given by a Kilis local implies, few exchange offices could also have the capacity to serve as a monetary deposit based on trust. The local townsman witnessed that a five square-meter small office could deliver his friend, whom he accompanied, a bulk of money after his friend gave a password to the shopkeepers. No exchange apparently had been made, but his friend received some money.

11 The elder brother indicated that the family had migrated from Iraq to Raqqa and then settled in this village in his grandfather's time and assumed their ethnic origin as Arab.

12 The rise in price (TL) of an ounce of gold (31,10 gr) every five years is as follows in the regulated domestic market: 6 in 1950; 9.25 in 1955; 16 in 1960; 14 in 1965; 21.50 in 1970; 80.50 in 1975; 1,835 in 1980 (Sağlam, 1991).

13 *Çarşıyı biz indirir kaldırırdık. Borca aldığım mal satılıncaya kadar ben birisinin kasasına para koydum. ... O biraz geç geldi. Gelene kadar kasayı mark dolarla doldurdum ya. Bir çuval mark dolar topladım Kapalıçarşı'da İstanbul'da.... Burdan altın götürürsün, ordan döviz alırsın.*

14 Drawing on his interviews with exchangers with a Kilis origin, Sağlam states that the traders of Kilis used to sell the gold to Jewish exchangers in Doğubank Office Block in exchange for foreign currency (1991: 63). My interviewees, on the other hand, only pointed to the Grand Bazaar. Contrasting these accounts, I assume that the exchangers in Doğubank performed their role until the mid-1960s before the Grand Bazaar gained significance.

15 *Biz terfi ettik, köyden şirketi buraya getirdik.*

16 Here I rely on my interview with Ata, a member of extended families with Syrian relatives from a border village.

References

Andreas, P. (2011) "Illicit Globalization: Myths, Misconceptions, and Historical Lessons", *Political Science Quarterly*, 126(3): 403–425.

Dündar, U. (2013) "Ürkütücü Gerçeği Açıklıyorum", *Sözcü*, 21.12.2013.

Erdoğdu, A. (2013) "Dostluğun Mimarı Berber Yaşar çıktı", *Birgün*, 23.12.2013.

Galemba, R. (2008) "Informal and Illicit Entrepreneurs: Fighting for a Place in the Neoliberal Economic Order", *Anthropology of Work Review*, 29(2): 19–25.

Galemba, R. (2012a) "Taking Contraband Seriously: Practicing "Legitimate Work" at the Mexico-Guatemala Border", *Anthropology of Work Review*, 33(1): 3–14.

Galemba, R. B. (2013) "Illegality and Invisibility at Margins and Borders", *Political and Legal Anthropology Review*, 36(2): 274–285.

Gates, C. (1989) *The Historical Role of Political Economy in the Development of Modern Lebanon*, Oxford: Center for Lebanese Studies.

Green, S. (2009) "Of Gold and Euros: Locating Value on the Greek-Turkish Border", *East-BordNet COST Action IS0803 Working Paper*, 1–23.

Hatip-Karasulu, H. A. (2005) *Making Sense of Mafia in Turkey: Conceptual Framework and a Preliminary Evaluation*, unpublished PhD thesis, Boğaziçi University.

Hiçyılmaz, S. (2013) "Özal'dan Erdoğan'a Hortum İstikrarı", *Evrensel*, 24.12. 2013.

Kahraman, A. (2013) "Berber Yaşar'ın Dönüşü", *Yeni Özgür Politika*, 21.12.2013.

Keyder, Ç. (1987) *State and Class in Turkey: A Study in Capitalist Development*, London: Verso.

Mercan, F. (2002) "Bankama izin verilseydi şimdi bir numaraydım", *Zaman*, 4.6.2002; Retrieved 18 September 2014 from http://arsiv.zaman.com.tr/2002/06/04/haberler/h12.htm.

Milliyet (1988) "Kilis "Küçük Sicilya": Babaların 13'ü Kilisli", *Milliyet*, 13.12.1988.

Naylor, R. (2004) *Wages of Crime: Black Markets, Illegal Finance, and the Underworld Economy*, Cornell University Press.

Nordstrom, C. (2000) "Shadows and Sovereigns", *Theory, Culture and Society*, 17(4): 35–54.

Öniş, Z. (2004) "Turgut Özal and his Economic Legacy: Turkish Neo-Liberalism in Critical Perspective", *Middle Eastern Studies*, 40(4): 113–134.

Öniş, Z. and Webb, S. B. (1994) "Turkey: Democratization and Adjustment from Above", pp. 128–184 in S. Haggard and S. B. Webb (eds.), *Voting for Reform: Democracy, Political Liberalization and Economic Adjustment*, New York: Published for the World Bank, Oxford University Press.

Roitman, Janet (2004c) "The Garrison-Entrepôt: A Mode of Governing in the Chad Basin", pp. 417–436 in A. Ong and S. J. Collier (eds.), *Global Assemblages: Technology, Politics, and Ethics as Anthropological Problems*, Hoboken, NJ: Wiley-Blackwell.

Sağlam, M. H. (1991) *Türkiye'de Altın Ticareti*, unpublished MA thesis, İstanbul University.

Tagliacozzo, E. (2007) "Thinking Marginally: Ethno-Historical Notes on the Nature of Smuggling in Human Societies", *Journal of the Canadian Historical Association*, 18(2): 144–163.

Tuşalp, E. (2005) "Otel Odası Pazarlamacıları", *Birgün*, 24.9.2005.

Yetkin, Ç. (1989) "Türk Mafyasının Kasası İsviçre 1–7, *Milliyet*, 28.8.1989 – 3.10.1989.

Yükseker, D. (2003) *Laleli-Moskova Mekiği: Kayıtdışı Ticaret ve Cinsiyet İlişkileri*, İletişim.

5 Kilis as "little Beirut", markets and illegality

This chapter details the shifts in the social and urban landscape as well as the scaling up of the local economy to the domestic market. It dwells upon the cultural mechanisms through which economic accumulation through illegal means could be socially accepted as wealth. I develop this argument by drawing on the community norms and values, including ethical principles and religious references that regulate and normalize illegal practices of trade. The transnational shadow economy shifts community norms and values, allowing the normalization of illegal accumulation as rightfully gained wealth as well as altering the meaning of border into a mere economic resource in the eyes of town dwellers, which led them to disregard strong kinship ties when their economic livelihood was threatened.

The last section gives an overview of the present condition of the shadow economy with reference to the local reactions against the transfer of Syrian guests from Hatay to the border camp in Kilis. It is suggested that the town lost its salience in the post-1980 period with the outmigration of new wealth. Nevertheless, despite declining profit margins, the dependency on cross-border trade has rendered the border gate even more salient. The local reactions indicate in what way the trope of 'border as gate to livelihood' organizes the narratives about cross-border trade as normal practices of everyday life, while reinforcing at the same time the social boundaries between local dwellers and Syrian migrants despite strong kinship bonds and close contact across the border. This case illustrates the moral legacy of the transnational shadow economy in Kilis town.

Shadow economy carving out the new urban and social landscape

As mentioned in the previous chapter, the Koyuncu family was not only involved in the gold trade but it was also engaged in contraband of several goods circulating from Beirut to the Turkish border. The family moved to the town and restored its business in an arcade shop, by selling the contraband goods coming from Beirut. The interviewees' recollections point to the entry of consumer goods along the borders, breaching the high tariff walls of the Turkish economy before the trade liberalization of the 1980s. The trade liberalization in Beirut had transformed the city into an "open market" facilitating the circulation of Western and Far Eastern exports into the inlands of southeastern Turkey. The Baathist subsidies also contributed to the reduction of price in exported goods and attracted Kilis traders to scale up their business.

The family began to bring contraband goods from Syrian Aleppo and sell them to the customers that visited the town from other cities. Mustafa revealed that while he appeared to be selling tableware at the front receiving custom, the rear side of the store remained as a coordinating point where his brothers ran the gold business and ensured its distribution. These goods were sold to domestic tourists who were driven to the town on bus tours for its famous contraband bazaar. The level of appeal that the town's shops reached led local residents to name Kilis "little Beirut," turning the town into an open market, even though these goods were deemed as *kaçak* when they crossed the town's boundaries. The customers had to be able to afford the risk of being caught by police patrols and being deprived of what they had. However, the goods could be freely sold and bought in the arcade shops that blossomed along the main axis of the town in the late 1960s.

The overflow of goods accelerated the introduction of the money economy and the transformation of the urban landscape, which irrevocably undermined the historic identity of the town that the old wealth was most attached to.[1] The first arcade, called Adalet Çarşısı, was built by the town municipality in 1967 on Cumhuriyet Street, the main avenue of the town dividing the old center, developed in a circular form into two hemispheres, and aimed at responding to the need to provide a marketplace for goods coming across the border and sold by street sellers at several corners.[2] Adalet Çarşısı accommodated 36 shops in the two-store building and started to bring in a yearly rental income of 115,000 Liras by the time the total trade of the town was estimated at 50 million Liras (Konyalı, 1968). The second arcade erected by a known smuggler and spurred astonishment among town dwellers according to an interviewee, since the owner could finance the construction of the arcade alone, that is with its financial resources affording what could be only afforded in that time by the municipal resources.

The introduction of the money economy also interrupted intergenerational occupational transmission by eradicating the old artisanship already in decline. The town had 75,092 residents in 1965, the rural and urban population being almost half.[3] In 1968, the trade already had a major share involving one-third of the population according to the records of the town's chamber of commerce, ranked after laboring in agriculture.[4] The industry was largely traditional, mostly consisting of small-scale factories and workshops for olive oil and grape molasses, as well as silk textiles with handloom weavers. The growth of contraband trade had an adverse effect on traditional artisanship, including soap making, stone masonry, tannery (*tabaklık-sepicilik*), sack making (*çuvalcılık*), coppersmithing, blacksmithing, saddle making (*saraçlık-palancılık*), rug weaving, shoemaking (*yemenicilik-köşkerlik*), plow making (*sabancılık*), wickerwork (*hasırcılık*), and crafts like radio and watch repairing which were already in decline. The interviewees' recollections indicate that the majority of craftsmen abandoned their vocations, and rented a shop in the newly constructed arcades in order to do trade.

The traditional trade-artisanal area, located on the southern hemisphere down Cumhuriyet Street, had an integrated structure with the old urban texture composed of bazaars, mosques, and bathhouses.[5] The building of Adalet Çarşısı on Cumhuriyet Street was a precursor to pulling trade life along the middle of the historic town center. In the following decade, more contraband goods hit the shops in the arcades that are lined up on the main street of the town. The old two-store stone houses with a courtyard located on the street were doomed to be demolished in order to be replaced by cement buildings. The number of shops in the arcades climbed up to 424.[6]

The border scholar Neşe Özgen (2005) remarks that the owners of arcades in Gaziantep and Kilis were local notables and merchants–moneylenders. The building of arcades was concomitant with the rapid urbanization, often accommodating the shops on the downstairs of a high-rise apartment block. The urban planning of 1967 accelerated the rapid urbanization as the road building works destroyed the historic texture by widening narrow alleys and streets and the succeeding municipal administrations sustained similar decisions of urban development by expropriating several monumental domestic dwellings for demolition (Bebekoğlu and Tektuna, 2008; also see Map 5.1). In terms of the familial meaning of housing, this meant the symbolic detachment of the lineage located at the paternal house

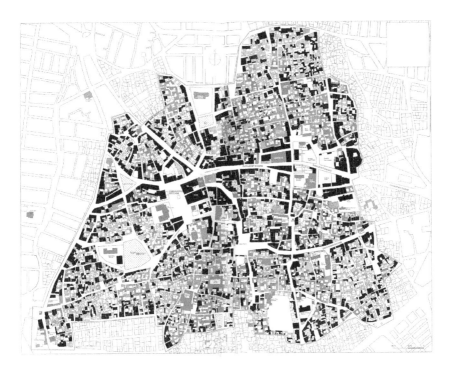

Map 5.1 Urbanization in the historic town center (adapted by the author from the map by Bebekoğlu and Tektuna, 1998). The dark grey colored plots show the concrete buildings, while the light grey ones are traditional dwellings. As it clearly illustrates, the concrete buildings are mainly concentrated along the main axis of the town, following the road building works with the urban planning of 1967.

and the old dwellings came to be "treated primarily as a form of investment, an economic patrimony whose transmission to younger generations is expected to be carried out through the sale of the house itself and its conversion into liquid capital" (Bertaux-Wiame and Thompson, 1997, p. 133).

My interview with the nephews of the abovementioned gold trader revealed that the urban development turned into a drift promoted by their neighbors, local merchants and even municipality employees. The trend was actually enhanced by the economic conditions of the 1970s under which high inflation forced the money earners to opt out of gold and foreign currency as a means of stabilizing value. The strict regulations of these markets led to the reckoning of rents to them and the sale and rental transactions for residential properties and arcade shops were made with foreign exchange and gold rather than national currency.

The custom of buying gold for newlyweds is a known cultural practice in Turkey as a wedding gift as well as a form of financial support offered in communal solidarity for helping the establishment of the new household. The gold signifies for people not only a desirable good but also money, even if worn on the skin (Strathern, 1975; cited in Green, 2009). The circulation of gold and foreign currency across the border made them more readily available in Kilis town and helped to shape the dwellers' lives and sense of value. The interview with the nephews, Behire and Tijen, exemplified the shift from paternalistic relationships based on traditional production to the money economy. The two sisters, whose father was an artisan, not only witnessed the enrichment of the family as their uncle was engaged in the gold trade but also actually watched over the gold traffic. The recollections of the elder sister stated that she was entrusted with the load of gold with her husband in order to put it out of sight before it had been shipped to the next destination. Unlike when women pinned a piece of gold onto her wedding dress, she could change the gold bars between her hands in order to pile them up in a corner of her house.

Each sister having their stone houses demolished for the construction of arcades could catch the recent trend towards the end of decade and regretted their delay in not having reaped the benefit of it. The local contraband trade almost stopped after being busted by the military raids after the 1980 coup and the shops were left idle or the rentals were not satisfying (See Figure 5.1). They had brilliant memories about how much the rental income of the shops was worth and that it should have been in gold. Whether they had been given by the building engineer an accurate floor space for their property occupied a significant part of the conversation between them during the interview. The conversations underlined their understanding of wealth and culminated in the confession of the elder sister that she would have been a smuggler as a profession if she were a man. This shift in the urban and social landscape of the town was clearly connected with the growth of the domestic underground market of gold and foreign currency as well as the informal sector, particularly in Istanbul.

Scale-up of shadow market from local to national

The local economy was increasingly enmeshed with longer-distance buyers in the domestic market. This was conditioned in two ways: either the potential buyers

Figure 5.1 Arcade shops left idle (Photo courtesy of the author)

or consumers themselves from different cities visited the town or Kilis traders established commercial links with these cities. A wide range of items crossing the border started to circulate in the domestic market: kitchen utensils with crystal glassware, porcelain tableware and assorted cutlery, high quality household electrical appliances and electronic goods including Phillips and Japanese-branded radios and tape players, perfumery and cosmetics goods, cigarette paper, playing cards, weave beads and other indispensable goods of daily life including watches and straps, eye glasses, lighters, nail clippers, and stationery.

The boom created by the arcades was dramatic enough to stir Mustafa into likening the town to Paris, as a source of inspiration for its nineteenth-century arcades. According to an interviewee, the town used to receive some 20 to 30 buses of domestic tourists thanks to its famous arcades in the 1970s. For Walter Benjamin, the Parisian Arcades constituted an important aspect of experiencing the urban public life of the century (Benjamin, 2002). The arcades of Kilis town were certainly incomparable to Paris's luminous, spacious, and elegant architecture. The arcades of the town also lacked the significant features highlighted by Benjamin: strolling, window-shopping, and observing. The customers were there on purpose and to buy contraband goods.

They did not even have the chance to observe the goods, if they wanted, for instance, to buy a small weapon. As a former shopkeeper said, when a potential customer demanding a small weapon came to the shop, he used to make the customer wait in order to ask another fellow townsman who is a known dealer. The fellow townsman took the pistol from a hidden place behind the racks where goods in sight were aligned. He returned back to the shop with the pistol and took the customer to a vacant place so that the customer could practise with the weapon.

The arcades of the town and the goods in them still promised a mass consumption experience that Benjamin outlined in his Arcades Project. The influx of foreign goods that were not available to the consumers in the formal domestic market or only available for high prices due to tariff walls forged the consumption experience into the passageways crowded by domestic tourists. From the viewpoint of consumers visiting the town, the goods in the shop racks seemed to embody the Western or Japanese industrial development and the level of quality, compared to 'shoddy' domestic goods. Border scholars demonstrated that the material objects circulating across the border acted as symbolic vehicles as they embodied national or transnational identifications,[7] a fact that should have fueled the demand to obtain 'foreign' goods for cheaper prices. Several interviewees, including Mustafa, said that they were involved in selling domestic merchandise on the arcade shops and the latter were paid higher prices than their real value because the customers surmised them to be foreign. The sales were good enough to make Mustafa earn on a gainful day an amount equivalent to the shop rental in the arcade. This could be the reason why the consumers visiting the town for its 'free contraband market' could easily be cheated by domestically-produced lower-quality or quality merchandise brought from Istanbul wholesalers and fraudulently sold to their customers.[8]

Kilis shopkeepers and traders not only brought merchandise from Istanbul, but they also supplied contraband goods to other cities, particularly the informal sector of Istanbul concentrated in Eminönü district. Mustafa's interview revealed that the family used to send contraband goods to Istanbul hidden in secret places in the car, bought by a trader that would sell these goods to a wholesaler and, eventually, these goods would be put on mobile stalls by the street sellers of Eminönü. For example, the profit range for the Schneider pens smuggled across the border and transported to Istanbul would be 20% after reducing the expenses, and the traders benefited from the competitive advantage of selling them for almost half the price when the same pens are legally exported and put to sale on the national market. Kilis traders could also invoice these goods in order to avoid a police investigation when necessary. They could ask the importer company, which sold the same good to domestic market, to officially make out an invoice in exchange for a certain price.

In the 1970s, the border gates and seaports gained prominence over the mined frontier. In 1975, the number of passport owners in the town reached the number of 27,000, almost one third of the population.[9] The signing of the trade agreement with Syria in 1974[10] increased the number of transport companies. The trucks shipped local agricultural products to Syria (pine nuts, pistachios, olive oil), consumer goods and parts (hoovers, electrical plugs for household appliances, chandelier glass), textile products (carpets, prayer rugs, fabric and garments), raw materials and semi-manufactured products (processed cattle intestine and catgut).[11] Still, these companies were implicated in the unregulated trade as the foreign exchange regime strictly controlled the entry of currency.[12] They were also suspected of smuggling goods on their way back, by hiding cigarettes in particular in secret places in the trucks.

Just like the gold trade, the contraband of consumer goods from Beirut became increasingly connected with the transnational shadow economy whose actors were large-scale traders. The early examples of fraudulent export, usually associated with the neoliberal period, were observed. The interviewees' recollections suggested that the entrepreneurs, including several from Kilis, produced the paperwork at the border customs for the transit shipping of goods from Beirut to Iran, except that the latter never crossed the border.

The goods circulated to Kilis and nearby cities as they were shipped by the small boats landing on the shores of the gulf of Alexandretta. There are numerous stories told by former smugglers about eluding the coastguard by buoying the contraband goods on the sea or, when caught, by simply getting rid of them by throwing into it. Nevertheless, the large bulk of these goods were smuggled into Turkey by ships docking at Mersin port and transported to cities like Istanbul. As early as 1970, the ratio of contraband to domestically-produced goods found in the local shops of the town was estimated to be about 20%.[13]

Morality of smuggling

Despite the civilian and state efforts for launching industrial enterprises, Kilis town failed to lead an industrial development during the import-substituting industrialization and it has lacked ever since state investments on industrialization.[14] The accounts introduced in this chapter suggest that the transnational shadow economy sustained in Kilis a distinctive redistributive mechanism largely controlled by large-scale entrepreneurs within a context where the state failed to promote public and local investments in industrialization and offered employment opportunities to the town community. Though illegal, town dwellers regard cross-border trade activities as legitimate. For instance the barber's 'success' is a standing promise for the middle and lower strata of Kilis town. His success not only consists of his moving up the social ladder by running an astute business strategy, but also of achieving the transformation of his image from a crime boss to a decent businessman, whose philanthropy is rewarded by highly-placed state officials with presented honor plates. His frequent visits to the town, appearance in local media coverage, and gatherings with members of the town authority invited to Istanbul for social dinners he hosts all give clues about the delicate balance between secrecy and transparency that the town locals try to keep. Adapting Galemba's words, while Kilis natives keep their businesses secret in order to avoid competition and the anger of higher level authorities not involved in the bribery network, they also maintain visibility to ensure community support and sustain the image of legitimate work (Galemba, 2012, p. 11).

Roitman argues that the unregulated trade and gang-based banditry designates an ethical realm where "one evaluates the nature of licit practice, as well as representations of the self and self-conduct" (2006, p. 265). The main point here is that the banditry activities cannot be conceived in terms of a juridical morality that distinguishes between right and wrong, good and bad. To put it differently, she suggests that the moral values that regulate these activities cannot be understood within the framework of universal principles about human nature. Roitman is guided here by the definition of ethics asserted by Michel Foucault as she contrasts it with the concept of morality. Foucault's notion of ethics refers to the subjectivation within the power relations, to "a mode of questioning about the self and the construction of the self in the world" (Roitman, 2006, p. 267). In the light of her arguments, I do not consider the economic practices of trade in Kilis as lying beyond the realm of justice and morality. These practices are governed by community norms and values. Following Roitman, I argue that these norms and values draw on local power structures, state regulations, code of smuggling and religious references, rather than moral principles.

My interviewees do not consider smuggling as a unitary category, but they distinguish it in terms of legitimate and illegitimate practices. For instance, an old generation woman from traditional landed notables suggests that even though Kilis locals were involved in smuggling gold and consumer goods in the past, it was fortunate that they did not engage in the heroin trade. Other interviewees include the small arms trade as an activity to avoid. My interview with a former smuggler reveals the comparison of smuggling with theft:

> In our view smuggling is an honorable crime, but you have to serve time in prison with men who committed infamous crime like adultry. You go without fear, defeat the military and earn your own money. You do not cheat somebody of his rights. Nobody mess with the smuggler, people makes him the representative of inmates. [Smugglers] take information from inside [prison administration]. He has not any slighest harm to someone. He has not any harm to the state. It is only a crime regarding taxation.[15]

The same interviewee also considers smuggling in terms of distinctions between helal (religiously lawful) and haram (religiously unlawful): "We used not to bring playing card because it is gamble and breaks up families. We used not to bring heroin and small arms because it was religiously unlawful and for fear that we would not do well".[16]

The distinction between helal and haram makes the religious reference clearly visible. But it is not possible to fix these categories either. For instance, when I visited few border villages in company with Hamit, a 26-year-old astute trader, doing cigarette and oil trade, we stopped in the village where my companion had grown up. We were hosted by my companion's family friend. Our host moved into conversation with my companion and complained about difficulties the border villagers experience and said: "Okey, [smuggling] is unlawful by religion. But when you are in it and you endure its difficulties so much, you come to the point of questioning in which way it is unlawful".[17] My young companion Hamit replied by repeating his father's words. For his father, an imam who used to join his fellow peasants in smuggling across the minefield, smuggling was not religiously unlawful but detested (*mekruh*). Mekruh is distinguished from religiously lawful and unlawful and it denotes the practice of soul (*nefis*) upon which the believer has to decide not to do it. So, these accounts suggest that smuggling should be considered as an ethical practice.

Lastly, I will mention about the codes of smuggling. The codes of smuggling in Kilis have again religious connotations. As stated in Chapter 2 on methodology with reference to my encounter with Urup Ismail, there is a "smugglers' sharia" that used to determine the codes of conduct regarding smuggling activities. Sharia was highly active during the import-substitution period. It also used to secure the perpetuation of smuggling activities by defining the rights and liabilities of individuals engaged in these activities. Why did the smugglers name their own codes as sharia? I believe that it was not related with their intention to make these codes as valid as the state laws are supposed to be. It probably refers to the decisiveness of resolutions made by sharia, as the saying that 'the finger cut off by sharia will not get hurt' suggests. As my encounter with Urup İsmail shows, I was not able to learn about sharia from its enforcers. But I had the idea that it was a guild-like organization recruiting the smugglers who proved themselves to be reliable.

The realm of unregulated trade practices puts forth the trust relationships because they lack a legal framework or because traders cannot rely on institutional state mechanisms (Yükseker, 2004; Atalay-Güneş, 2012). Smugglers in Kilis had to count on trust relationships in doing trade with Aleppo merchants and among themselves. The harsh punishments forestalled the prevention of behaviours harming the trust relations among smugglers such as denouncement of smuggling activity to law enforcement, breaking one's word and cheating. At the same time, sharia compensated for the loss of partners or the injuries of peasants working for their boss. An interview with a retired teacher who had knowledge of sharia explained its role as follows:

> [The rights of] smugglers who invest in share of a joint business should be protected. For example, when he is shot by a bullet, his treatment in the hospital. [Sharia] will tell the smugglers who have taken this injured man to the job to take care of his family. Sharia used to gather and make its call in order to prevent victimhood.[18]

My interviewee also emphasized that sharia offered patronage to the lower stata by using its relationships with local bureaucrats and politicians.

These examples highlight which practices are considered as legitimate livelihood strategies and which values and norms support them. The meaning of smuggling in Kilis town does not originate in the transgression of law but it stems from its organization at the community level. The distinctions about legitimate/ illegitimate practices are not moral statements because they do not judge it right or wrong. But they are ethical principles that regulate the realm of illegality and the smugglers are expected to follow them.

Shadow economy and order of things at the border

As the Syrian conflict grew and the migrants began to pile up at the Turkish border, the Disaster and Emergency Management (AFAD) governed by the Turkish

Prime Ministry decided in early 2012 the shift of Syrian refugees in Hatay to Kilis town and a camping ground at the zero point to the border ordained to be built in order to 'host' them. The news was quickly heard in Kilis and met by the local 'civil society' with discontent. The local representatives of opposition parties and semi-governmental trade and agricultural bodies, as well as the local transit shipping and passenger transportation companies and drivers, were present at a meeting with the governor and mayor in order to express the civil society's discontent about the coming of Syrian migrants.[19] They pointed out that the border gate constituted the main source of livelihood in the town. Their complaint was that the indwelling of rebels against the Syrian government so near to the border would incite Assad's anger and prompt him to close the gate. For town dwellers, keeping the border open was vital in order to sustain the local economy.

Actually, their discontent was shared by the broader community of shopkeepers who struggled to earn a living by small-scale contraband across the border despite the decreasing profit margins particularly in the last decade under the fragile conditions created by the global economy. Several reasons were speculated about: any attack to the border camp committed by a lunatic supporter of the Syrian regime or PKK fighters deterred from crossing to the Turkish side of the border in this region until now could bring Turkey to the brink of war. Moreover, lending an ear to the rumours circulating across Hatay, the town dwellers suggested that the Syrian migrants were indulged in stealing and prostitution and their coming was of no good. However, for the middle strata of the town, the main concern was economic. The transportation and trade and retail sector composed the lion's share of income distribution.[20] Not only did the shipping and transportation companies yield revenues from the flow of goods and passengers across the border, but shopkeepers also benefited from the price differentials between the two countries.

Still the means of livelihood mainly depended on the illegal trade across the border associated with the informal sector. According to a local entrepreneur, there were about 15 registered companies in the town licensed for international transport, usually carrying the load of Gaziantep factories to the Middle East. These companies were implicated in the contraband of goods that would be worth of carrying in the hidden places (*zula*) of their vehicles like cigarettes, automobile spare parts, and drugs on their way back to Turkey. But, the majority of the traffic was due to the registered passenger taxis as well as civilian cars which work as unregistered companies of families by putting several cars alternately for the use of border crossing in order to bring goods from Syria.[21]

The drivers usually drove to Syria with empty tanks so that they could fill up with cheaper Syrian gasoline and immediately sell the unconsumed remainder. They could carry and hide contraband goods or without declaring them at customs, thanks to the complicity of officers. Not only the expensive items, but also Ceylon tea, Syrian chips, and chocolates, cheap housewares, and trifles, Chinese-made electrical appliances and imported underwear, cosmetics and perfumes of poor quality were put to sale in the store racks, next to the better quality goods so that the middle-class public employees and university students could buy what they needed among the offered options. None of these purchases would be taxed. The unregistered jewelry–currency exchange offices concentrated at a district in the town center complemented this untaxed economy.

Nevertheless, all these economic practices formed the 'normal order' for the town dwellers. The illegality of smuggling goods and circumventing the customs law was regarded as socially acceptable. Cigarettes, especially, carried to the nearby city of Gaziantep could be seized by the gendarmerie or police who stop cars on suspicion of contraband. Yet, the cigarettes were sold everywhere in the town, at the stores or stalls of itinerant street traders who used to wait at their usual stop and in most of the grocer's stores in the neighborhoods with no obstruction. As you were stepping out of your apartment, there was a chance of running across an errant boy on his motorcycle with a used sport bag between his legs frequenting the neighborhood grocers to ask whether his smuggled cigarettes were needed. The town dwellers rightfully questioned the selling of contraband goods within the town with ease, while these items were treated as contraband and seized at the road blocks by the police patrols outside the provincial boundaries. The demand of town dwellers was that the door to their means of living should not be closed. That is, they expected that the local authorities should overlook the illegal trade, albeit without sanctioning it.

Thus, the local reactions to the worrisome decision of transferring Syrian migrants to the town could make sense with reference to the trope of border as gate to livelihood. The middle strata's fear about the closing of the border gate—in Turkish, the word "gate" is the same as "door"—implied their concern with losing access to the means of earning income from illegal cross-border trade, promising to yield some extra gains both for dwellers as well as custom officers. The border dwellers classified all sorts of economic practices in the town as "smuggling," whether it be informal or illegal, large-scale or small-scale, crossing goods from the customs gate or across the minefields at the border. When the border dwellers declared their concern about the closing of the border gate, they actually meant the whole borderland.

Kapı (the door) in Turkish figurative speech also means *ekmek kapısı* (literally, the door to bread), the place where one can earn an income or a living. Thus when the border dwellers suggest that the border gate is their place of earning a living, they actually make a strong claim that no one should mess with their bread. They consider smuggling to be a business, and being a smuggler a job. As an owner of a transport company and a spokesperson of the initiative of company owners suggested, every region had "its own boon" and in Kilis, being a border zone, the dwellers had the right to make the most of it, i.e. make use of the privileges and benefits of smuggling. The dwellers asked for the maintenance of the status quo because, they believe, this is what they are used to. Hence, the local reactions to the incoming of Syrian migrants gives insights into the mechanisms through which economic and political stability could be sustained so far, though underlying at the same time how economic interests, cultural affinities, and political agendas were delicately lined up in a geography of tensions.

Concluding remarks

In this part, I have focused on the economic, cultural, as well as political mechanisms by which the economic accumulation through illegal means provides upward mobility and becomes socially accepted as rightfully earned wealth in Kilis town. I have discussed in what ways the transnational shadow economy created a redistributive logic providing employment and altered the social stratification and occupational structure during the import-substituting period. The illegal trade activities involved the participation of notable families, large-scale entrepreneurs, extended families with a rural background, artisans, and border villagers.

As the case of Kilis suggests, contrary to the public convictions that the neoliberal regime promotes corruption by undermining moral values, economic accumulation through illegal means is a morally laden realm regulated by ethical principles and rules of conduct. It may also have religious connotations. On the other hand, the redistributive logic of illegal trade was also limited since the economic accumulation tended to create family monopolies and establish connections with transnational organized crime. The story of the Koyuncu family reveals the significance of local power structures in terms of patronage and kinship in facilitating monopoly tendencies. So, to what extent did smuggling, as town dwellers term it, constitute a 'moral economy,' as they manipulated and circumvented the state regulations of the gold and foreign exchange market and import tariffs in order to capitalize on the price differentials, rates of foreign exchange and demand structures? The following chapters address this question with reference to the urban and rural poor strata. They discuss the ways that patronage and kinship ease the participation of the poor in illegal cross-border trade.

Notes

1 For the historic identity of the town, see Bebekoğlu and Tektuna (2008); Taşçıoğlu (2013).
2 My interviewee İhsan used to be employed as a worker at the directorate of technical works at the municipality in those years and, thus, had first-hand knowledge.
3 Urban population: 38,095; rural population: 36,997 according to the 1965 Population Census by Turkstats.
4 In Konyalı (1968) the records for other sectors is as follows: agriculture 30%, trade 28%, artisanship 26%, service sector 2% and industry 14%. The year for the records is not stated.
5 An axis that vertically parted from Cumhuriyet Street along the road to Sabah Pazarı (a fixed wet market), where the biggest old covered bazaar demolished in the mid-1930s was located, accommodated the artisanal shops (Akdemir and İncili, 2013). Thus it is understood that the old trade and industrial area was developed around the big covered bazaar and stretched out to the southeast, that is, in the direction of Odunpazarı Street and Hasanbey Bathhouse. In the present day, the latter spot still bears the traces with the shops of ironsmiths, bladesmiths, tinsmiths, and a traditional shoemaker. The interviewees' recollections indicate the location between the Grand Mosque and Tuğlu Bathhouse as another marketplace, which was called little bazaar. The town square where the old Tekye mosque was located at its southern corner could also accommodate a marketplace with stalls of goods.

6 The number is taken from 1990 State statistics, cited in Ökkeş (1994). I assume that the number had not been changed since the late 1970s because any of the buildings was not demolished after the 1980 coup. Nevertheless, Mustafa stated the number of shops as 700 to 800. He might have included the shops outside the arcades.

7 See Reeves (2007); Yükseker (2007); cf. Pelkmans (2006). For Yükseker, the flow of signs associated with Western values accompanied the flow of goods from Turkey through informal trade and helped their entry to the nascent capitalist markets of the former Soviet Republic in the 1990s. On the contrary, Pelkmans' ethnography of the Turkish–Georgian border explores how the cheap and flimsy Turkish goods flowing into Ajaria contributed to the redrawing of boundaries between "us" and "them." Pelkmans calls these markets "treacherous" because they embodied "the disillusionment with the capitalist change and the massive influx of new consumer goods" (Pelkmans, 2006, p. 172).

8 For example, Alaattin, a shopkeeper in a partly deserted arcade, said that they used to sell domestic goods rather than smuggled ones. Alaattin also recounted how the contraband goods could easily be sold in the arcade shops as the shopkeepers negotiated bribes with the law enforcement officers.

9 1977 Turkish National Grand Assembly Research Report dated on January 28, 1977 of the Parliamentary Comittee no. 10/14 on the cleaning of minefields and distribution of land among landless peasants; see TBMM, 1977. The town population was exactly 92,759 dwellers.

10 The first trade agreement signed on September 17, 1974 designated the goods allowed for exportation and importation until the end of 1982. For details see the Economic and Technical Cooperation Agreement between Turkey and Syria published in Official Gazette No. 17785, 17.8.1982.

11 The information about trade with Syria largely draws on my interviews with Mahmut, a young member of a traditional landed family owning an international taxi company; Yasin, a middle-class tradesman dealing with white sale and wholesale trade; and Murtaza, a young transport company owner carrying the goods of industrialists in Gaziantep to Middle Eastern countries.

12 The interview with the former head of Kilis Chamber of Commerce and Industry emphasized for example the Decree No. 17 regarding the Law No. 1567 on the Protection of the Value of Turkish Currency. The resolution requires the import of value received in exports within three months and its exchange at a Turkish bank within 10 days. See Decree no. 17 issued by Cabinet Decision 6/763 and published in Official Gazette on August 11, 1962.

13 See the statements by the Minister of Customs and Monopoly Ahmet İhsan Birincioğlu in TBMM, 1970. However, deputies taking the floor remind that the contraband trade in volume tends to increase in the last decade and harms the Turkish economy a great deal.

14 A significant attempt for industrial investment in the town was a public-private joint venture of an oil factory in the mid-1950s (Çolakoğlu, 1995). The project fell behind and was cancelled by the succeeding government. The only state investment in the town was an alcohol (*suma*) factory of Tekel built in 1944 and privatized in 2004. The development thrust consisted of modernizing agriculture under the auspices of the Marshall Plan in the 1950s. According to 2011 data, there are 30 industrial enterprises in the free industrial zone and 250 workplaces—mostly motor mechanics—in a small industrial site (Taşkesen, Erke, Bölükbaşı and Karipçin, 2011)

15 *Bizim görüşümüzce kaçakçılık şerefli bir suç ama zina gibi yüz kızartıcı suçlar işleyen adamlarla yatıyorsun. Korkmadan gidiyor, askeri tepeliyor, kendi paranı kazanıyorsun. Kimsenin hakkını yemiyorsun. Kaçakçıya kimse dokunmuyor, onu koğuş sorumlusu yapıyorlar. İçerden bilgi alıyor. En ufak kimseye zararı yok. Devlete de zararı yok, sadece vergi açısından bir suç.*

16   *Kağıt kumar, ev yıkıyor diye iskambil kağıdı getirmezdik. Dinen haram, işimiz rast gitmez diye uyuşturucu, silah getirmezdik.*
17   *Tamam [kaçakçılık] haram da, işin içinde olup bu kadar çilesini, kahrını çekince de nesi haram diyorsun.*
18   *Kaçakçılıktan hisse yiyenlerin [hakkı korunur]. Mesela kaçağa giderken mermi yemiş, onun hastanede bakımı. Onu işe götürenlere bu adama hastanede bakacaksın, ailesine bakacaksın derler. Bu gibi mağduriyeti önlemek için şeriatlar olurdu.*
19   See the details of the four-hour meeting in *Gazete Kilis* (2012). The civil society organizations of the town also met the protests of Islamist non-governmental organizations (NGOs) in solidarity with Syrian opposition that took place in the town in May and July 2011 with harsh reactions and resilience. My interview with Bahri, the speaker of the May demonstration, a lawyer and volunteer of the Humanitarian Relief Foundation (İHH) in Kilis, was manhandled by the drivers of the passenger transportation companies rallying against the solidary protests.
20   In 2000, the gross domestic product of the town was distributed as trade (37%), agriculture (32.8%), and industry (10.8%) (Bayraktar, 2003). According to the 2003 Data of Kilis Chamber of Trade and Industry, 30% of total registered companies (counted as 1,004) work on food production and trade, while 10% operate in the shipping and transportation sector. However, the latter dropped in the last decade.
21   The interview with a taxi company owner revealed that there were about 11 to 12 registered companies in total. But there are more unregistered companies than registered since the applications to the authorization certificate for passenger transportation was first ceased and then annually limited by 5% of the total number of registered companies in the previous year (see the circular of the Ministry of Transport on December 29, 2006, no B.11.0.KUG.0.10.00.02/275-31997). The companies have to compete with Syrian taxis, allegedly paying less for registration in their country.

References

Akdemir, İ. O. and İncili, Ö. F. (2013) "Şehir Morfolojisi ve İktisadi Yapi İlişkileri: XIX. Yüzyıl Kilis Şehri Örneği", *Kilis 7 Aralık Üniversitesi Sosyal Bilimler Dergisi*, 3(6): 79–100.

Atalay-Güneş, Z. N. (2012) *Theorizing 'Trust' in the Economic Field in the Era of Neoliberalism: The Perspectives of Entrepreneurs in Mardin*, unpublished PhD thesis, METU, Ankara.

Bayraktar, F. (2003) *Kilis İli Uygun Yatırım Alanları Araştırması*, Ankara: Türkiye Kalkınma Bankası Araştırma Müdürlüğü.

Bebekoğlu, S. and Tektuna, M. (2008) *Kilis kültür envanteri: kentsel mimari, kırsal mimari, yazıt eserleri*, Kilis Valiliği.

Benjamin, W. (2002) *The Arcades Project*, R. Tiedemann (ed.), H. Eiland and K. McLaughlin (trans.), New York: Belknap Press.

Bertaux-Wiame, I. and Thompson, P. (1997) "The Familial Meaning of Housing in Social Rootedness and Mobility: Britain and France", pp. 124–182 in D. Berteaux and P. Thompson (eds.), *Pathways to Social Class: A Qualitative Approach to Social Mobility*, Oxford: Clarendon Press.

Çolakoğlu, Ş. (1995) *Kilis Tarihi Üzerine Deneme*, Ankara: Kilis Kültür Derneği Yayınları.

Galemba, R. (2012) "Taking Contraband Seriously: Practicing "Legitimate Work" at the Mexico-Guatemala Border", *Anthropology of Work Review*, 33(1): 3–14.

Gazete Kilis (2012) "Kilisliler Suriyeli Mülteci İstemiyor", 5.1.2012; Retrieved 17 May 2013 from www.gazetekilis.com/kilisliler-suriyeli-multecileri-istemiyor.

Green, S. (2009) "Of Gold and Euros: Locating Value on the Greek-Turkish Border", *East-BordNet COST Action IS0803 Working Paper*, 1–23.

Keyder, Ç. (1987) *State and Class in Turkey: A Study in Capitalist Development*, London: Verso.

Konyalı, İ. H. (1968) *Âbideleri ve kitâbeleri ile Kilis tarihi*, Kilis Belediyesi.

Özgen, Neşe (2005) "Sınırın İktisadi Antropolojisi: Suriye ve Irak Sınırlarında İki Kasaba", pp. 100–129 in B. Kümbetoglu and H. Birkalan-Gedik (eds.), *Gelenekten Geleceğe Antropoloji*, Epsilon Yayınları.

Pelkmans, M. (2006) *Defending the Border: Identity, Religion, and Modernity in the Republic of Georgia (Culture and Society after Socialism)*, Ithaca, NY: University of Cornell Press.

Reeves, M. (2007) "Unstable objects: corpses, checkpoints and "chessboard borders" in the Ferghana valley", *Anthropology of East Europe Review*, 25(1): 72–84.

Roitman, Janet (2006) "The Ethics of Illegality in the Chad Basin", pp. 247–272 in J. Comaroff J. and J. Comaroff (eds.), *Law and Disorder in the Postcolony*, Chicago: University of Chicago Press.

Strathern, M. (1975) *No Money on Our Skins: Hagen Migrants in Port Moresby*, Port Moresby: New Guinea Research Unit Australian National University.

Taşçıoğlu, S. (2013) *Tarihi Kentlerde Kimlik Sorunu: Kilis Örneği*, unpublished MA thesis, Mustafa Kemal University, Hatay, Turkey.

Taşkesen, H., Erke, F., Bölükbaşı, İ. H. and Karipçin, S. (2011) *Kilis İl Çevre Durum Raporu*, Kilis: Kilis Valiliği Çevre ve Şehircilik İl Müdürlüğü.

TBMM (1970) TBMM Tutanakları (Proceedings of TGNA), 28th legislative term, volume 5, 3rd period, 1st session, 26.5.1970.

TBMM (1977) TBMM Tutanakları (Proceedings of TNGA), 4th period, 4th session, 1977.

Yükseker, Deniz (2004) "Trust and Gender in a Transnational Market: The Public Culture of Laleli, Istanbul", *Public Culture*, 16(1): 47–65.

Yükseker, Deniz (2007) "Shuttling Goods, Weaving Consumer Tastes: Informal Trade between Turkey and Russia", *International Journal of Urban and Regional Research*, 31(1): 60–72.

Part III
Rural and urban poor

6 Peasantry turning into border laborers

This chapter provides a historical analysis of the illegal trade from the viewpoint of the lower strata as it reveals the practices and meanings that they invoked for normalizing their engagement in illegal practices. The urban and rural poor found in the unregulated trade the means to emancipate themselves from the paternal relationships that were based on large landholding and kinship during the import-substituting industrialization (ISI) period. However, the paternalistic domination was replaced by the patronage of large-scale entrepreneurs, reaping the profits of unregulated trade by hiring the poor as porters. Yet, the economic boom in the 1970s helped the latter to make money and establish themselves briefly as independent "patrons," despite the dominance of large-scale entrepreneurs, who acted as local dealers collecting the goods from the small-scale traders and distributing to the domestic market.

The cross-border trade regulations introduced in the mid-1990s revived the transnational shadow economy and drew the poor despite declining profit margins and arbitrary risks within the context of neoliberal policies, which transformed a secure salary into a privilege rather than a right, undermined the peasantry and created a welfare regime based on assistance dependency and political patronage. The chapter shows that the kinship relations proved to be resilient and versatile ways, allowing cross-border alliance among families and normalizing the exchange practices, deemed illegal by local authorities.

Independent peasantry in Kilis

As discussed in Chapter 2 under the heading "Question of notables as early capitalists," the large landholding was not conducive to the commercialization of agriculture in Kilis town. However, the economic patronage of landlords was so strong that they dominated town life until the land tenure did not sustain their class position any longer. Not only the peasants but also the small merchants were dependent on the landlords for the on-credit dealings that were made in the time of harvest (Altuğ, 2002, p. 83). I argue that the decline of paternalist relations in agrarian production is crucial to understanding what ways living conditions of lower strata had changed. Since town life was basically organized by agrarian economy, I extend my discussion of the previous chapter to the question of peasant status and explore it by drawing on oral interviews with rural dwellers. Despite the theoretical controversy over the transition to capitalism in the agrarian sector, the scholarly literature agrees on the increase of petty-commodity production in agriculture after World War II (Keyder, 1989). Yet it should be noted that the shifts in the land tenure and peasant status in southeastern Anatolia are highly debatable. Did the sharecropping agreements completely dissolve in the town and had the peasantry been emancipated from their dependency on the landlords on the basis of petty-commodity production?

The agrarian studies emphasized the complex processes after the World War II by which the mechanization of agricultural production and the rural migration to the cities led to the decrease of sharecropping arrangements between the landlords and the peasants (Karadağ, 2005). It is also argued that the mechanization of agricultural production helped to make the petty-commodity production among the peasants dominant. Çağlar Keyder suggested that the consolidation of state control on land against the seizure attempts by the landlords and the land reform in 1945 helped the peasants to enlarge their landholding, with the incoming of tractors that opened new fields to agriculture (Keyder, 1989). This tendency was counterposed by the growing technical capacity of large-scale landholders with tractors, pushing the peasants off the land and transforming them into seasonal and wage labor or simply making them unemployed. The Passavant regime promoted the vested interests of *eşraf* families on their land left in Syria and ensured their peasants' dependency on landlord-managed estates as sharecroppers, laboring in return for modest provisions usually paid in kind and for protection under patronage relations. If expelled from their lands, the peasants were forced to provide wage labor for the landlords for salaries barely sufficient to subsist on. My interviewees Hayrullah and Şükriye, an old couple at a border village in the lowlands of Kilis (see Figure 6.1), described it as follows:

> We used to go weeding for 2,5 liras. We worked all day, sticking three dry breads under our armpits. If you can cook pilaf, you are like *agha*s and *pashas*. Soup is rare, it is a strong meal.[1]

As Keyder claims, the peasant landholding constituted an exception in this region: the landlords supported by the political power largely maintained their proprietorship on land and the petty-commodity production remained limited (Keyder, 1989, p. 733). In the case of Kilis town, it is plausible to suggest that the mechanization of agriculture did not lead to the domination of petty production, but rather its coexistence with sharecropping as well as wage labor.

The sharecropping in Kilis established the contract between the landlord and the tenant peasants as a patron–client relationship, which often comprised beyond mere economic interest. The landlord lent seeds or livestock in exchange for unpaid labor of the client peasant and the harvested produce was shared between them. The contract was usually based on sharing the produce in halves, a legal arrangement that was justified by the Ottoman manorial system: "You take the grain from the stack and you divide in halves."[2] Nevertheless, the landlord tended

Figure 6.1 Mudbrick houses in a lowland border village (Photo courtesy of the author)

to hold a stronger position when the landlord lent the seeds or engaged in usury by giving a loan to the peasant. The interviewees indicated that the peasants' obligations to the landlord were not merely economic:

> You cannot build a house unless the *agha* allows you to build it. You plant a tree and you become a proprietor. Even the house does not belong to you. If he [the *agha*] says to move, you move.[3]

Keyder argues that the sharecropping was already constrained by the early 1960s with the southeastern region populated by Kurdish sedentarized tribes (Keyder, 1989, p. 733). Kilis geography accommodated a significant number of Kurdish as well as Arab and Turcoman tribes. But the interviewees' recollections indicate that the sharecropping arrangements were not only dominant among the populations governed within the tribal economic organization, but were also relevant in the estates managed by the descendants of Ottoman military-bureaucrats and ulama families. The interviewees recalled that they crossed in the early 1960s through the Passavant gates to work on the landlord's estate in Syria: "We used to go to Syria for harvesting grapes in exchange for a basket of grape. All [the land] belonged to the *agha*."[4]

Keyder acknowledges, on the other hand, that in cases where the sharecropping arrangements coexisted with petty-production, the mechanization process made the opening up of new lands possible and helped the sharecroppers to establish themselves as independent peasants. The coexistence of landlord-managed estates and peasant lands would generate tension between the landlords tending to enclose new land and the peasants striving to establish their full rights of possession on their new and former fields. Karadağ's study on local notables of Gaziantep reveals stories about the belligerence of peasants in the 1960s from the viewpoint of landed families that were having a hard time understanding why the peasants wanted to reclaim lands from them (Karadağ, 2005, p. 117).

Drawing on this scholarly debate, I argue that the dissolution of sharecropping arrangements was facilitated in Kilis town by two factors. First, the demarcation of the border impaired the ties that Kurdish *aghas*[5] and other landlords had with their estates by rendering them absentee and enabled the sharecropper peasants to reclaim the lands under the new political context of the 1960s. As an interview with the *mukhtar* of an eastern border village demonstrated, the landless Arab peasants could also own land by making informal proprietorship agreements with the absentee landlords to buy the land for nominal rates. The lands of absentee landlords at the Turkish side of the border could not be rented or sold pursuant to the retaliation in kind made against the Syrian government that had confiscated the estates of Turkish citizens left in Syria after the land reform of 1958. Second, as mentioned earlier, the pattern of inheritance did not allow the possession of large landholding among the heirs of notables. As the mechanization was not conducive to the commercialization of agricultural production either, the management of estates did not remain as an option for the reproduction of class position among the heirs and they tended to sell their plots.

The data about the land tenure in Kilis show that the large landholding remained significant even as late as 1990, with 6% of the landholder families owning one third of the total arable land (Kesici, 1994, p. 136). Despite the controversial status of proprietorship on land, these data imply that the mechanization of agricultural production in the 1960s did not change the structure of land tenure dramatically, though it led to an increase in small proprietorship among the peasants. In 1990, half of the landholder families only held a plot of a size equal or below 0.5 hectares, with the average plot size being 0.1 hectare for 112 villages of Kilis. The small size of the plots indicated that the farming household could not benefit from the mechanization to its full potential and had to divide its forces among several plots in case it owned more than one. Following Köymen's discussion on self-sufficient peasantry in the 1930s, the size of the land sufficient for a rural household to subsist should be at least 0.7 to 1 hectares (Köymen, 2009, p. 29). Thus, land with such small sizes would not be enough for rural families in Kilis to compensate for their labor.

The fact that the outmigration in Kilis villages started in the 1950s actually supports this view. Rural migration consisted of the movement from the countryside to the town center and to a nearby city, Gaziantep, where migrants sought employment opportunities in urban settings, particularly in the second half of the 1960s. Also, the peasants could offer their labor seasonally, moving every year to the fertile plains of the Hatay-Maraş rift valley for cotton and pepper hoeing and harvesting.[6] According to Keyder, the migration pattern in rural Turkey did not require the small proprietors to sell their plots and they tended to rent them out to the middle-sized owners. In this way, they could maintain the ownership of the land while leaving their villages seasonally or permanently. The landless peasants, in turn, were likely to lease the land of small proprietors. Thus, these tenancy arrangements were dissimilar to those made between the landlords and peasants, where the former benefited from a specific economic patronage and domination.

It can be concluded that the peasantry of Kilis achieved an independent status vis-a-vis the market relations, as they were emancipated from their dependency on the paternalistic relations of agricultural production in the 1960s. But the small proprietorship of land remained limited or insufficient for the subsistence of rural families. Hence, they were driven to the shadow economy that had been growing alongside the newly emerging wealth accumulation strategies. Like the dwellers of border villages, rural migrants in the town sought to benefit from the income-generating activities linked with contraband trade. The increasing ratio of urban to rural population in Kilis as of the mid-1960s and the acceleration of the migration wave in the following decade demonstrate that the shadow economy boosted the economic prospects for rural families (Kesici, 1994). But the shadow economy quickly developed new dependency relations for the rural and urban poor, drawing on the traditional paternalistic and patronage relations.

Border economy and new relations of dependency

Hüseyin the porter was *agha*. Can you imagine a porter turned *agha*?[7]

Mevlüt seemed astonished that these words came out of his mouth. He forgot me and lost himself in talking to Cemal next to him, also present for my research interview. Cemal, a retired teacher, helped me by asking Mevlüt for his participation in the interview. As I sought to interview the border dwellers about their involvement in the growing shadow economy during the ISI period, Cemal suggested that Mevlüt would be a good source to listen to since he used to work as a piggybacker for the large-scale entrepreneurs. He could give first-hand information about the patronage relations which the poor had been pulled into. In fact, his statement about Hüseyin the porter was a clear illustration that the border did not only function as a means of upward mobility, but also shifted the power relations in the region. The high profit margins raised the stakes on the border and prompted the official as well as unofficial figures of authority to control the illegal crossings.

The unregulated trade in Kilis border dated back to the consolidation of the border. According to the minister of internal affairs Şükrü Kaya, the high numbers of border-crosser peasants that went to work on the Syrian fields in plough and harvest times implicated them in practices of 'smuggling' (BCA, 1931). The peasants tended to smuggle goods on the belt wrapped around their waist or on their packsacks. Kaya was also convinced at the end of his investigations that there were individuals mingling with the laboring peasants but employing themselves in smuggling. The interviewees' recollections also supported Kaya's observation that the Passavant regime of border crossing enabled the growth of contraband as early as the 1930s.[8] Hayrullah and Şükriye, the old peasant couple from a lowland border village, said that they used to pay a small bribe to the guards of Passavant gates in order to have clearance without submitting a permit and labored in the fields across the border, while sneaking a few contraband goods on the way back. The payoff for their labor was so low and usually in kind. So the contraband could yield extra income for the rural families. But the poor ones were unlikely to engage in large-scale trade and their practices basically consisted of exchange of their own products for their basic needs.

Research on post-colonial borders in Africa illustrates that local exchange could evolve within the context of colonial history, while the colonial powers tolerated the porosity of borders with financial disincentives rather than severe measures of criminalization for the sake of bringing the region under their control as regional economies were penetrated by merchant capital (Donnan and Wilson, 1999). Studies on post-colonial borders often neglect the repercussions of the colonial past on border communities, but they still reveal a history of control and regulation by dwellers over border traffic and the emergence of a common identity, encompassing border communities by drawing on their residential claims and everyday struggle to fight against economic marginality and insecurity (Flynn, 1997; Doevenspeck and Mwanabiningo, 2012; Galemba, 2012a). Flynn, for instance, argues that a sense of territorialization could be entrenched among border communities within the context of exchange across the border (Flynn, 1997).

Similarly, the colonial domination in Syria during the interwar period was likely to allow the growth of shadow markets in Aleppo provinces. While the Turkish authorities were determined to thwart smuggling practices across the border, reports from the region suggested that the French Mandate was reluctant to settle down the disturbances caused by local dwellers' trespassing of the border.[9] The demarcation of the border, customs tariffs, and the imposition of the French monetary system impaired inland trade by weakening the merchant class in Northern Syria (Bouchair, 1986). But Aleppo still continued to receive agricultural products and livestock from the Euphrates basin. The interviewees' recollections indicate that big flocks of sheep and cattle and olive oil were imported by Kilis smugglers to French-Mandate Syria.[10] As mentioned in Chapter 3, the notables having their estates across the border circumvented the Passavant regime in order to put their agricultural produce in the Syrian market. Thus, the agricultural produce of Kilis was smuggled to Syria as well. In turn, the high tariff barriers imposed by the French Mandate made specific goods such as coffee, sugar, gas oil, European fabric, and British tailored jackets go on the black market in Kilis town.

As mentioned in the previous chapter on the emergence of new wealth and the middle strata, family monopolies emerged to control the unregulated trade insofar as the shadow markets for contraband produced more lucrative businesses for them. The traditional paternalistic relationships on land tenure also supported these family monopolies. Kilis had been a frontier zone edging the Arab provinces of the Ottoman Empire, where a large confederacy of tribes were governed as chiefdoms. For instance, both the *Ekrad* of Okçu İzzeddinli (Kurds of Okçu İzzeddinli) and İlbeyli Turcomans of Aleppo organized into territorial entities of sanjaq-chiefdoms rather than simple lineage systems and could contain ethnically or religiously different segments of both nomadic and resident tribes (Soyudoğan,

2005; Öztürk, 2005).[11] As Martin van Bruinessen discussed about the persistence of Kurdish tribes in the modern Middle East, the administrative centralization since the late Ottoman period had a diminutive effect on large tribal confederacies as they turned eventually into large and more homogenous tribes, and large tribes turned into smaller ones (Bruinessen, 2002, p. 7). The Ottoman policy of resettlement in the sixteenth and seventeenth century disaggregated the Turcoman, Arabic, and Kurdish nomadic tribes notoriously known for banditry or accused of committing it in Aintab and Kilis as well as broader areas of Syria, Iraq, and other Kurdish emirates by expelling them to Arab or western provinces like Rakka and Adana, though they kept returning back. But, the tax farming system persisting till the nineteenth century in the region secured the power of tribal chieftains, who tended to impose their rule as district governors. Thus, they often had to confront the central Ottoman administration.

In the late Ottoman and early Republican era, we find the status of tribal chieftains degraded in Kilis town and ranked after the ulama and dynasty descendants. But few still retained their power as *agha* among the local community and their extended families, with a loyal clientele around them. They could capitalize on the unregulated flows across the border, as illustrated by the case of Mustafa's family. As a reminder, Mustafa's father was likely to be a leading lineage of an Arab tribal chieftainship resettled to Rakka (where it moved from Iraq according to İsmail, another family member), from where they migrated to Kilis. At least, his family genealogy aligned with my discussion here.

The border people may have the advantage of living at the edge of two differential legal and economic systems. Border studies document their interstitial power to benefit themselves in controlling and regulating cross-border movement (Flynn, 1997). Their locatedness at the border granted them the force to impose an unofficial toll for the non-locals or distant-range entrepreneurs and to serve them as guide through the minefields or mountain passes shielded from view and mediate between them and local customs/gendarmerie officers by providing brokerage services in negotiating the bribe. They were called *vasıta* (mediator) in Kilis town, according to Mevlüt and Cemal. Nevertheless, few studies focus on the ways in which the cross-border ties are enmeshed within local power structures, kinship, and community norms.

Bruinessen rightfully indicated that the tribal chieftains were best placed to conclude such profit-sharing arrangements without being apprehended at once (Bruinessen, 2002, p. 16). My interview with Vahab, an old-generation *agha*, who served as mayor in the mid-1980s and a parliamentary deputy in the early 1990s, revealed that the chieftains could guarantee economic security and access to wealth not only for their lineage, but also for local people through clientelist relationships. The local rumours implied that the chieftain retained his power through an involvement in contraband. In the light of interviewees' recollections, the local power figures in the town engaging in the border economy of contraband comprised not only the tribal chieftains, but also village headmen (mukhtars, locally called *kiya*). The mukhtars had a privileged place in making profit-arrangements with the military or acquiring an informal toll from the contraband road crossing through their villages.

As Janet Roitman discussed about the gang-based road banditry at the tri-bordered Chad Basin, the emergence of such 'traditional' power figures caused the pluralization of regulatory authority over cross-border movements recognized as legitimate, whether official or unofficial, and they became the final arbiters of employment and enrichment for local dwellers (Roitman, 2004b, p. 426; see also Roitman, 2004a). She developed her argument against the backdrop of African economies that were immersed into the world economy in the neoliberal era. However, the unofficial figures of authority in Kilis town tended to employ political patronage relations for their clientele and developed more intermingled relationships with the state politics by actually taking part in it within the context of the shift to neoliberalization policies. As the chieftain said about his serving as mayor in the post-military coup period, he sponsored the wage payments of municipality workers by drawing on his own funding, while the municipality was the major employer in the post-military coup context of budget shortage in the public sector and a sharp decline in contraband.

A study on the cross-border ties between Palestinian Bedouins settled in self-governing enclaves and Israeli-occupied territories reveals that the border crossings were organized as an informal border economy motivated by profit-seeking and facilitated by border-crossing workers, smugglers, employers, and state institutions (Parizot, 2014). The border economy not only refers to new types of economic activities peculiar to the border but it also points to an "industry of crossing" (Hernandez-Leon, 2008; cited in Parizot, 2014) where formal and informal entrepreneurs as well as state officials are involved in the process of regulating cross-border movements. The growth of transnational flows forged border economies in which all parties involved struggled to increase their shares within the context of profit-sharing arrangements, the shares particularly rising as the border grew more impermeable. The mining of the Turkish–Syrian border from the mid-1950s onwards raised the stakes of the border economy and turned it into a complex organization.

As the border economy in Kilis became more lucrative, it brought out figures whose power stemmed from their role in making profit-sharing arrangements or in guaranteeing their maintenance on behalf of the official border patrolling agents. *Emanetçi* (trustee) and *muhbir* (informant) were two such figures pointed to perform their role in a more complicated border economy. *Muhbir* reported the border traffic that went unnoticed by the military posts and yielded a share from the contraband goods seized by the gendarme. But they also brokered profit-arrangements between large-scale traders and gendarme officers and they actually took part in the contraband convoys crossing the border. *Emanetçi* was entrusted with the bribe money by the gendarme officers until they left the town since they could not deposit it in a bank or have it present. Neşe Özgen wrote that the *emanetçi* people could prosper to the degree of owning an arcade store in Gaziantep and Kilis (Özgen, 2005, p. 21). Mustafa's interview informed that these figures were possibly preferred by the gendarme officers based on trust and confidentiality relationships already established by the former profit-sharing arrangements for the crossing of contraband among them. In Mustafa's words, "they [the military] took their savings from us. We used to help them [in keeping their money] in case that they are caught."[12]

On the other hand, the border economy also helped the ascendancy of new figures of authority that mimicked and contested at the same time the traditional paternalistic power figures.[13] These figures established their authority as *kiya* or *agha*, as they were enriched through contraband although they did not possess the characteristics of kiya or they did not descend from a tribal chieftain. *Kiya* was a traditional paternalistic figure inherited from the early Republican times when the local administrations lacked legal framework or public budget. The interviewees' recollections hinted that the *kiya* was likely to be the tribal chieftain or head of wealthy families who could establish their authority due to the patronage relations with their client peasants and afford financing the needs of the village or hosting village guests. Thus, the *kiya* was expected to possess the features of notables with recognition, respect, and generosity, needed for developing paternalistic relationships of dependency over their clients. However, the fact that the contraband enabled the rise of smugglers from a poorer background into figures of authority led to the creation of new relations of dependency in which the enriched smugglers had made themselves elected as *kiya* or assumed the position of *agha*.

The rise of smugglers from a poor or rural background among the ranks of local notables meant the undermining of structural constraints thwarting upward mobility among different strata, as detailed in Chapter 3 on local notables. The interviews with the lower strata revealed the hatred among them against the old stratified structure based on large landholding. For instance, there were stories about the anonymous wrecking of the ancient tombs belonging to the notable families in the old cemetery of the town before its resettlement to the outskirts or demolition of the notables' houses by the 'populist' municipality as a sign of disfavoring the notable strata. These stories acted to bring out the hatred or discontent of the lower strata against the notable families. Thus, the rise of new figures of authority could be considered as the transfer of local power that both contested and mimicked the traditional paternal relationships.

Mevlüt recalls that the smuggler *agha* established a patronage relationship with the piggybackers who carried his load across the minefields, which was reminiscent of the relationship between the landlord and sharecropper. The piggybacker who worked for the *agha* could be defined as not only his employee but also a helper that was responsible for running errands for him, making his barbecue, and waiting on his table while he was sipping his *raki* alone or in the company of his guests. The *agha* in return provided protection for his helper by covering the expenses in case of injury or taking care of his family during his imprisonment. These patronage relations offered the poor border dwellers that endangered themselves by sneaking into the minefields security against military patrols or detonation of undetected mines, which could be landed by the gendarme officer or rival smugglers in cases of disagreement. Still most piggybackers had to undertake the risk of mine detonation even though the smugglers' convoy was led by a guide across the minefield. They usually treaded a footpath around the military post with which an agreement was settled (Özgen, 2005). But the lack of patronage could expose them to larger risks of injury, as the salience of former smugglers with prosthetic legs among my interviewees indicated.

Whether under patronage or not, the poor border dwellers were likely to carry the goods ordered by longer distant-range buyers. The growth of contraband emancipated the border peasants from dependency relationships based on the traditional organization of agricultural production. But, as Neşe Özgen indicated in her discussion about the political economy of Syrian and Iraqi borders, it also led to the proletarianization among the lower strata (Özgen, 2005). The poor dwellers were incorporated in the unregulated trade as labourers and entangled in new patronage relationships with the large-scale traders. As the salience of the border gate increased in the mid-1970s, small-scale trading allowed the poor to make money, which allowed, in their point of view, them to establish themselves as "patrons." For example, Asiye, a 78-year-old peddler that used to go trading in the 1970s with her husband, recalled the thriftless spending of her husband at Aleppian casinos: "I am the patron and my husband is spending the money." A border villager assumed his position as patron in his interview, as he grew up to hire a group of piggypackers to smuggle goods across the minefields for long-distance buyers. These patrons, however, were doomed to fail in climbing up the social ladder as they usually lost their winnings or they were cheated by some adversaries. The military coup of 1980 impeded unregulated trade and led to an increase in migration from rural areas, though the cross-border trade regulations introduced in the mid-1990s tempted the poor anew.

Cross-border trade: blurring the boundaries of legal and illegal

I consider cross-border trade as the epitome of the blurred boundaries between legal and illegal realms, characterizing the neoliberal regime of accumulation. The cross-border trade regulation was initially introduced in 1985 in Ağrı within the context of the neoliberal urge to promote the inflow of foreign currency and oil commerce with Iran, which had been underway since the late 1970s (Öztürk, 2006). The trade regulation was stretched out to the eastern and southeastern borders of Turkey before long and it underwent several modifications in the following decades. Cross-border trade was a special trade regime, allowing a limited volume and value of trade on the basis of local needs and population of the border city. It excluded items of international export and import and transit trade, while banning the marketing of imported goods nationwide. In other words, the traders could only sell their imported goods in the local market within the city district.

However, cross-border trade boosted the transnational shadow economy along Turkish borders in the form of profiteering by large-scale entrepreneurs as well as politicians and bureaucrats. It could be argued that the Özalist principle of "economic punishment for economic crime" (Öniş, 2004) allowed the blurring of boundaries between legal and illegal in the neoliberal era, as it offered rent-seeking enterprises room to maneuver. On the other hand, the replacing of prison sentences with penalty fines enabled the urban and rural poor of Kilis town to be freed of their bonds of dependency to the large-scale entrepreneurs, while rendering them more vulnerable to growing debts as they could not afford the pecuniary penalties and losing their goods to the custom officers.

The complicity of the state officials in facilitating unregulated activities and border traffic and the neoliberal state offering a legal structure to these activities further complicates the subject under study. Though seemingly paradoxical, while the neoliberal state provided a legal framework for unregulated trade practices, it did not alter the fact that "they are still deemed either formally illegal or based on fraud" (Roitman, 2004a). I suggest that the dominant discourse on cross-border trade fails to recognize it within the framework of neoliberal governance and the shadow economy. The international congress on cross-border trade held in Kilis and sponsored by the governorship and the Chamber of Trade and Commerce in 2010 is exemplary.[14] The congress compiled numerous papers that put the blame on the misuse of regulations by traders. The papers adopted the state's notion of illegality and assumed an ideal practice of cross-border trade by stating the problems in its implementation. As Abraham and van Schendel would argue, these studies failed to recognize that there could be no such universally shared definition of illegality from which corrupt behaviours deviated (Abraham and van Schendel, 2005, p. 8). These scholars indicated that multiple competing authorities interact at the border zones and the illegal status of circulating goods was constantly renegotiated.

The research drawing on border studies analyzed the implementation and practices of formal cross-border trade by situating its perspective on border dynamics. Their analyses offered the room to observe multiple contesting authorities involved in the distribution of 'illegal' profits (Kolay, 2012), the cultural processes promulgating 'illegal' practices of transnational economy as legitimate (Özgen, 2005, 2007), and the construction of 'official' boundaries for circulating goods as a product of multiple and contingent judiciary acts informed by political, economic, and social processes (Bozçalı, 2010, 2014). The arrangement of oil commerce into cross-border trade enabled its shift into a pipeline on wheels in the 1990s, allowing long-distance truckers to carry and sell excess oil in their tank. The high percentage of taxes on oil (and cigarettes) levied by the Turkish state had made it a remunerative one. Two pipelines on wheels, built with the Iraqi Kurdistan and Azerbaijan, particularly promoted cross-border trade as a lucrative business as the Turkish state permitted the drivers, including the busmen and local commuters, to carry oil in oversized tanks until this practice was abolished in 2001. The state eventually decided completely to put an end to the oil trade at the turn of the millennium.[15]

The research on the oil trade showed that the state not only regulated it, but also legalized the smuggling by readjusting the limits that the drivers could carry (Bozçalı, 2010; Özgen, 2005). It was argued that the legal framework that enforced or ended cross-border trade was often illegible because the state enacted it with memorandums and directives (Bozçalı, 2014). At the Iraqi border, the state opted to compensate for the loss of trade with Iraq after the Gulf War with oil trade and breached the United Nations (UN) sanctions by allowing the truck drivers to sell the excess fuel in their tank within the framework of international human aid delivered to Iraq. The arrangement also promoted the governorships as regulatory authorities, where the truck drivers had to apply in order get a license. In 1995, the Şırnak governorship benefited from its regulatory power to establish a foundation and levy tax from the crossing truck drivers.

In 1999, the Turkish Petroleum International Company was established to force the truck drivers to sell their oil and monopolize the marketing of oil coming from Iraqi Kurdistan nationwide. However, the cross-border trade regulations banned the sale of imported goods outside the city boundaries. The company also included the Şırnak deputy of the ruling party and the brother of the ex-mayor as partners and thus brought politics into the oil trade (Kolay, 2012, p. 142). All these developments did not stop, however, these practices being officially deemed as illegal. The parliamentary debates, official reports, and speeches delivered to the public recorded the oil trade period as causing remarkable economic losses due to smuggling or its sale on the informal markets.[16] Still, it was acknowledged that the definition of 'smuggled' oil was not easy to clarify. A parliamentary investigation on oil contraband in 2005 stated that:

> the definition of fuel oil entering along the borders with tax exemption or reduction resulting from certain local and regional tax immunity allowed by the governmental decrees concerning cross-border trade would come to mean pushing the limits of legal definition.[17]

Cross-border trade was reframed in the 2000s by a shift in the security perception of the state and its anti-corruption agenda. The decade of the 2000s saw consecutive amendments and modifications in the legal framework of cross-border trade and anti-smuggling. After the ending of oil commerce in 2001, new decrees and communiqués regulated the cross-border trade, which basically reiterated the clauses of abated regulations. The major difference consisted of the addendum and alterations that were taken as precautions against the misuse previously experienced. For instance, fuel oil—as well as other goods that were previously subject to smuggling—was excluded from the items that could be traded and the regulatory authority was centralized by replacing the governorships with the undersecretary of customs.[18]

It was argued that the ending of oil commerce originated in Turkey's discontent that the Kurdish regional government in Iraq benefited from oil proceeds (Kolay, 2012). Bozçalı's ethnography in local courts on Van exposed the fact that the local judiciary system could be manipulated by political motives. He detected a dominant conviction among the judiciary personnel that the oil (and cigarette) trade financed terrorism related to PKK (Bozçalı, 2014). Bozçalı also suggested that the same conviction dominated the government, setting off a series of amendments concerning the 50-year-old anti-smuggling law during the period 2003–2008. In some cases, the cross-border trade practices could be interrupted by police operations and governmental probes under the rubric of fighting against corruption that prevailed after the heightening of relations with the European Union (EU) and the International Monetary Fund (IMF). This was particularly true for Kilis.

The interviews revealed that cross-border trade lived its heyday after the 1996 regulation was legislated.[19] The regulation permitted the town dwellers to benefit from a specific customs regime in which the traders of both sides could import or export a certain amount of goods with reduced taxes. Cross-border trade quickly turned into a lucrative business for the region, allowing the urban poor as well as the middle strata to engage in bringing as many goods as they could carry in their bags or car trunks and yield profit due to price differentials. However, a notorious anti-corruption operation against border customs in the summer of 2000, called Parachute, interrupted the trade.

Hence, the regulations appeared as an employment opportunity or a side income for the town dwellers. The Chamber blamed the poor legal infrastructure for the malfeasance. According to the former general secretary of the Chamber, the cross-border trade regulations were not based on mutual agreement between Turkey and Syria anyway. He indicated that the formally regulated trade rules were violated by special directives and confidential circulars of the public authorities as well as transgressed by the traders taking advantage of the legal "loopholes":

> There was not actually a cross-border trade agreement reciprocally signed. Turkey enforced it unilaterally by issuing cross-border document and permit for the compliance to quota. We have told at the time that this would be a problem.[20]

After its interruption for a decade, the amendments made in the cross-border regulation to prevent the malfeasance drastically diminished the profit margins of unregulated trade, though they did not exclude it as a business option for local dwellers. While the Turkish government enacted the regulations unilaterally without reciprocation, the protective economy of Syria had not ever allowed a reciprocal agreement on cross-border trade. The signing of the free trade agreement between the countries, ratified in 2007 and forestalling a gradual remitting of all custom taxes, ruled out cross-border trade as a feasible option for Syria all together. Thus, the Turkish government continues to maintain the cross-border trade regulation in effect, though no local trader in Kilis has ever since made a trade license application according to the Chamber.

Concluding remarks

In this chapter, it is argued that the dependency of the lower strata on the paternalistic relationships of agricultural production was replaced by new dependencies based on community and kinship norms and connections across the border. While the sharecropping arrangements between the landlords and peasants declined in rural areas, the petty commodity production in agriculture remained limited and pushed the peasantry to urban settlements. Still, the rural poor could yield income from unregulated trade, though they were largely incorporated into the growing shadow economy as laborers employed by large-scale entrepreneurs. The increasing profit margins created an industry of crossing at the border, which was controlled by official as well as unofficial authority figures.

In the previous chapter, I raised the question about what ways illegal economic practices, constituting a redistributive mechanism and simultaneously giving way to monopolistic tendencies, could be regarded as a moral economy. Here I elaborated further on this question by highlighting a few aspects with an exclusive focus on the everyday strategies that the poor rely on in coming to grips with new border contexts. In this chapter, it is demonstrated that the poor dwellers eventually emancipated themselves from the patronage relationships of large-scale entrepreneurs at the expense of losing their protection against law enforcement at the border.

Notes

1 *2,5 liraya yolmaya giderdik. Bütün gün çalışırız. 3 kuru ekmek koltuğumuzun altına. Pilav yaptığın zaman ağasın, paşasın. Şorba bulunmaz, kuvvetli yemek.*
2 *Kömeden alır bideriyi, ortadan bölersin.*
3 *Ağadan izin almadan ev yapamazsın. Ağaç dikersen mülk sahibi olursun. Ev bile senin değil. Göç derse göç.*
4 *Suriyeye bağ kesmeye giderdik bir zembil üzüme. Hep ağa malıydı.*
5 The archive sources indicated that Okçu İzzeddin district (known as *Ekrad-ı İzzeddinli*) of Aleppo, where Kurdish tribal leaders ruled as *mir-i liva* (the district governor) until the late Ottoman era, was left within the border of French Mandate Syria with the demarcation of the Turkish–Syrian frontier. Some sources suggested that the word *Ekrad* (the plural of Kurd in Arabic) implied the Turcoman nomadic tribes living in Kilis region under the rule of Şeyh İzzeddin Begh in the sixteenth century (Akis, 2002, 2004; Öztürk, 2005), while the Kurdish historiography assumes that Ekrad-ı İzzeddinli referred to Kurdish tribes. See for instance the letter of Kurdish tribal leader Şeyh İsmailzade Hacı Hannan Agha, addressing the presidency of the Grand National Assembly in 1922 and asking for consideration of their inclusion along the Turkish border as the Turkish government negotiated with French Mandate the revisions of the newly delineated border (Bayrak, 2012).
6 Kesici states in his research dated 1990 that some 800 rural families, that is above 4,000 members of the population, were involved in seasonal migration (Kesici, 1994, p. 62). He further remarks that rural families also used to seasonally migrate to these plains in order to put their livestock out to pasture.
7 *Hamal Hüseyin ağaydı. Hamaldan ağa mı olur?*
8 Here I draw on my interview with İlyas, an old villager man, living at a border village in a one-story comfortable concrete house with a well-groomed lawn in his yard, an indication of his upward mobility as a piggybacker.
9 See the 1923 report of the Gaziantep governorship detailing the measures to be taken against smuggling cited in Öğüt (2010, p. 35) at footnote 39. Altuğ and White (2009) argued that the Turkish government and French Mandate intentionally overlooked the tribal banditry in order to capitalize on the instability of the border for extending their political influence beyond it.
10 For instance, Zeynep, an elder woman from an extended family which used to live at a border village, remembers vividly how her father and uncles used to smuggle livestock to Syria.
11 Though the word *ekrad* is the plural form of Kurd in Arabic, its ethnic implication is controversial. The nationalist historiography argues the *ekrad* of Kilis were Turcoman (Öztürk, 2005; Akis, 2002). Soyudoğan's study suggests that Okçu İzzeddinli tribes were basically Kurdish nomads, but it was not unusual for the tribes organized into a chiefdom to include or assimilate different ethnic groups, particularly because the latter

sought shelter among other tribes when they escaped from resettlement places (Soyu-doğan, 2005, p. 90). It is very likely that Okçu İzzeddinli tribes included Turcoman lineages as well.

12 *Askerler birikmiş parasını yanımızda alır. Belki tutulur diye biz ona yardım da ederdik.*

13 I am indebted for this idea to Ömer Özcan (2014) whose study on Kurdish smugglers in Yüksekova, the southeastern corner of Turkey, suggests that the unofficial road blocks by Kurdish peasants in order to collect a toll from the crossing smugglers could be interpreted both as repudiation and mimicry of the state sovereignty. Galemba inter-prets in the same vein the erection of tollbooths by the communities at the Guatemalan–Mexican border as mimetically reproducing the relation between territory, power, and identity (Galemba, 2012b).

14 See the proceedings e-book *1. Uluslararası Sınır Ticareti Kongresi Bildiriler Kitabı* published by Kilis 7 Aralık University; available at http://iibf.kilis.edu.tr/bordertrade/files/Kongre_kitapcigi.pdf

15 Özgen (2012) argued that the unregulated oil trade along the Turkish borders was not extinguished but changed form after 2003, following the US invasion of Iraq.

16 The State Minister Tunca Toskay held a press conference in 2002 announcing that the allowed quotas of fuel oil carried across Habur and Nahcıvan gates were exceeded many times more than official limits and the smuggling could not be prevented. He also added that the oil trade should be banned; see "Motorine sınır ticareti yasağı", *Milliyet*, August 16, 2002. The Minister of Financial Affairs Kürşat Tüzman spoke in the parliamentary assembly that cross-border trade diverted from its original purpose after the inclusion of fuel oil trade, which led to a tax loss of 3.5 billion USD during the period 1997–1999; see TBMM (2003, p. 219).

17 See, p. 36 in the report on "Research and determination of measures to be taken against the harms of oil smuggling to human and environment health" prepared by the par-liamentary committee on June 16, 2005 with no. 10/238. According to the report, all transactions of fuel oil were legal, but it was untaxed and placed on the informal mar-ket. It also pointed to the overlapping of legal and illegal by reporting the fraudulent trade as "mixing chemicals to the legally circulating oil in order to dilute it and raise the profit margins."

18 See Anti-Smuggling Law No. 5607 published in Official Gazette No. 26479 on March 31, 2007. Retrieved January 23, 2018 from www.resmigazete.gov.tr/eskiler/2007/03/20070331-1.htm.

19 The 1996 regulation sanctioned 13 border cities as well as their neighboring districts to enjoy the provision of cross-border trade. This regulation was modified in 1998 by abandoning the clause of "neighboring city." See the cabinet decrees on December 26, 1996 with no. 96/9025 and on June 4, 1998 with no. 98/11160.

20 *Aslında Suriye'yle karşılıklı imzalanmış bir sınır ticareti anlaşması yoktu. Türkiye tek taraflı olarak sınır ticaret belgesi ve kotaya uygun izin yazısı düzenleyerek bunu gerçekleştirmişti. Biz söylemiştik o zaman sorun çıkar diye.*

References

Abraham, I. and van Schendel, W. (2005) "Introduction: The Making of Illicitness" pp. 1–37 in I. Abraham and W. van Schendel (eds.), *Illicit Flows and Criminal Things: States, Borders, and the Other Side of Globalization*, Bloomington: Indiana University Press.

Akis, M. (2002) *16. Yüzyılda Kilis ve Azez Sancağında İktisadi ve Sosyal Hayat*, unpub-lished PhD thesis, Ankara University, Ankara.

Akis, M. (2004) "Tahrir Defterlerine Göre 16. Yüzyılda Kilis Sancağındaki Aşiretlerin İdareleri, Nüfusları ve Yaşamları", *Tarih Araştırmaları Dergisi*, 22(35): 9–31.

134   *Rural and urban poor*

Altuğ, S. (2002) *Between Colonial and National Dominations: Antioch under the French Mandate (1920–1939)*, unpublished MA thesis, Boğaziçi University, Istanbul.

Altuğ, S. and White, B. T. (2009) "Frontières et pouvoir d'État. La frontière turco-syrienne dans les années 1920 et 1930", *Vingtième siècle*, 103: 91–104.

Bayrak, M. (2012) "Suriye Kürtleri ve 1922 Mutabelatı", *Kürt Tarihi Dergisi*, 3: 48–55.

BCA (1931) Başbakanlık Cumhuriyet Arşivleri (The State Archives of the Prime Minister's Office), 180/244/6, 5.12.1931.

Bouchair, N. (1986) *The merchant and moneylending class of Syria under the French Mandate, 1920–1946*, unpublished PhD thesis, Georgetown University.

Bozçalı, F. (2010) The Illegal Oil Trade Along Turkey's Borders, *MERIP*, 41(261).

Bozçalı, F. (2014) "Hukuki-Maddi Bir Kategori Olarak Sınır: Türkiye-İran Sınırında Kaçakçılık, Mahkeme Süreçleri ve Sınırın 'Resmi' Temsilleri, *Toplum ve Bilim*, 131: 135–162.

Doevenspeck M. and Mwanabiningo, N. M. (2012) "Navigating Uncertainty: Observations from the Congo-Rwanda Border", pp. 85–106 in B. Bruns and J. Miggelbrink (eds.), *Subverting Borders: Doing Research on Smuggling and Small-Scale Trade*, Wiesbaden: VS Verlag für Sozialwissenschaften.

Donnan, H. and Wilson, T. (1999) Frontiers of Identity, Nation and State, Oxford, New York: Bloomsbury.

Flynn, Donna, K. (1997) "We Are the Border": Identity, Exchange, and the State along the Benin-Nigeria Border", *American Ethnologist*, 24(2): 311–330.

Galemba, R. B. (2012a) "Taking Contraband Seriously: Practicing 'Legitimate Work' at the Mexico-Guatemala Border", *Anthropology of Work Review*, 33(1): 3–14.

Galemba, R. B. (2012b) "Corn Is Food, not Contraband": The Right to 'Free Trade' at the Mexico-Guatemala Border", *American Ethnologist*, 38: 716–734.

Hernandez-Leon Ruben. 2008. *Metropolitan Migrants*, Berkeley: UCP.

Karadağ, M. (2005) *Class, Gender and Reproduction: Exploration of Change in a Turkish City*, unpublished PhD thesis, University of Essex, 2005.

Kesici, Ö. (1994) *Kilis Yöresinin Coğrafyası*, Ankara: Kilis Kültür Derneği Yayınları No. 12.

Keyder, Ç. (1989) "Social structure and the labour market in Turkish agriculture", *International Labour Review*, 128(6): 731–744.

Kolay, G. (2012) "An Economy of Survival and Reinventing the Way of Life: The Case of Oil Commerce in Southeast Turkey" pp. 129–146 in B. Bruns and J. Miggelbrink (eds.), *Subverting Borders: Doing Research on Smuggling and Small-Scale Trade*, Wiesbaden: VS Verlag für Sozialwissenschaften.

Köymen, O. (2009) "Kapitalizm ve Köylülük: Ağalar – Üretenler – Patronlar", *Mülkiye*, 35(262): 25–39.

Öğüt, Tahir (2010) "Milli Sınırların Oluşum Sürecinde Güneydoğu Anadolu'da Kaçakçılık Sorunu İle Dahiliye Vekili Şükrü Kaya Raporu ve Değerlendirilmesi", *1. Uluslararası Sınır Ticareti Kongresi Bildiriler Kitabı*, İktisadi ve İdari Bilimler Fakültesi, Kilis 7 Aralık Üniversitesi, 4–6 Kasım.

Öniş, Ziya (2004) "Turgut Özal and his Economic Legacy: Turkish Neo-Liberalism in Critical Perspective", *Middle Eastern Studies*, 40(4): 113–134.

Özcan, Ö. (2014) "Yüksekova'da Sınır Deneyimleri: Bir "Sınır Kaçakçılığı" Hikâyesi ve Barış Süreci", *Toplum ve Bilim*, 131: 162–185.

Özgen, N. (2005) "Sınırın İktisadi Antropolojisi: Suriye ve Irak Sınırlarında İki Kasaba" pp. 100–129 in B. Kümbetoglu and H. Birkalan-Gedik (eds.), *Gelenekten Geleceğe Antropoloji*, Epsilon Yayınları.

Özgen, N. (2007) "Öteki'nin Kadını: Beden ve Milliyetçi Politikalar", *Türkiye'de Feminist Yaklaşımlar*, no. 2. Retrieved from www.feministyaklasimlar.org/sayi-02-subat-2007/otekinin-kadini-beden-ve-milliyetci-politikalar/.

Özgen, N. (2012) "Kaçakçılığın ve Sınırın Öyküsü", Bianet, 14.1.2012. Retrieved 7 August 2014 from https://m.bianet.org/bianet/insan-haklari/135445-kacakciligin-ve-sinirin-oykusu.

Öztürk, M. (2005) "İzziye Kazasının Kuruluşu ve Milli Mücadeledeki Yeri", *Ankara Üniversitesi Dil ve Tarih-Coğrafya Fakültesi Tarih Araştımaları Dergisi*, 37: 29–45.

Öztürk, N. (2006) "Türkiye'de Sınır Ticaretinin Gelişimi, Ekonomik Etkileri, Karşılaşılan Sorunlar ve Çözüm Önerileri", *ZKÜ Sosyal Bilimler Dergisi*, 2(3): 107–127.

Parizot, C. (2014) "An Undocumented Economy of Control: Workers, Smugglers and State Authorities in Southern Israel/Palestine", pp. 93–112 in L. Anteby-Yemini, V. Baby-Collin and S. Mazzella (eds.), *Borders, Mobilities and Migrations Perspectives from the Mediterranean 19–21st Century*, Brussels: Peter Lang.

Roitman, Janet (2004a) "Introduction", pp. 1–22 in *Fiscal Disobedience: An Anthropology of Economic Regulation in Central Africa*, Princeton University Press.

Roitman, J. (2004b) "The Garrison-Entrepôt: A Mode of Governing in the Chad Basin", pp. 417–436 in A. Ong and S. J. Collier (eds.), *Global Assemblages: Technology, Politics, and Ethics as Anthropological Problems*, Malden and Oxford: Wiley-Blackwell.

Soyudoğan, M. (2005) *Tribal Banditry in Ottoman Ayntab (1690–1730)*, unpublished MA thesis, Bilkent University.

TBMM (2003) TBMM Tutanakları (Proceedings of TGNA), 24th assembly, 1st session, 21.1.2003.

van Bruinessen, M. (2002) "Kurds, States and Tribes", pp. 165–183 in F. A. Jabar and H. Dawod (eds.), *Tribes and Power: Nationalism and Ethnicity in the Middle East*, London: Saqi.

7 "The border gate will not ever be closed"

Livelihoods, aspirations, and reciprocity among the poor

The chapter explores the subjective perceptions and evaluations of the poor underlying their engagement in unregulated trade in order to reveal the "cultures" of unregulated trade. It demonstrates that the poor embraced unlawful practices "to create what they see as a worthwhile life, or a way of life worth living and fighting for" (Galemba, 2012, p. 8) rather than the urge to rise to wealth or make money hand over fist. I suggest that the cultures of unregulated trade were not only dictated by an economic rationality, but they also drew on community and kinship norms and habits. The poor creatively appropriated the governmental policies that sought to regulate or ban these trade relations and hegemonic discourses that marginalized their involvement in these activities. However, they also gave consent to the status quo, by assimilating that the means to their living were unlawful and by normalizing their unequal access. Rather than rejecting it, they chose to capitalize on it in order to yield differential profit as the stronger law enforcement at the border increases trade revenues. Furthermore, the remittent profiteering by large-scale entrepreneurs kept the prospect of upward mobility alive for the poor.

Border laboring in the neoliberal order

In February 2012 when the popular upheaval entered a second year in Syria, the conflicts between local opposition forces and the Syrian army were heightened and pervaded into the 7 km away Syrian border town Azaz. Syrian troops engaged in heavy gunfight and bombing against the town. As the clashes of arms broke out at the opposite side of the border, I expected that the drivers and peddlers crossing the border for small-scale trading would cease to commute across it. In fact, though declining, the daily cross-border commuting continued despite the successive deaths of two small-scale traders and a transport company driver from Kilis in crossfire in the opposite Syrian town. The killing of a 63-year-old woman peddler illustrated that the urban poor kept shuttling across the border to make a living out of it no matter what happened. A condolence visit to the house of the deceased informed about the social profile of the border commuters.

In a poor neighborhood on the outskirts, the relatives and neighbors of the deceased told me that she turned a deaf ear to her son's preaching to his mother not to go. Mother of six children, she wanted to cover the lack of home fixtures for her elder son who was preparing to be married and earn some pocket money for her daughter who was studying in a Western Anatolian university according to the female community in the house. Heading to Aleppo, she would sell the grape leaves she picked up from the vineyard to her acquaintances and buy from the Syrian local market any goods that she could sell door-to-door with a low profit margin in Kilis neighborhoods or she would deliver to an arcade shop in return for payment—known as 'hire' (*kira*) by the locals, since these commuters did not own the goods but they were simply hired for carrying them.

The deceased woman typically exemplified the women who had to earn a living as they were widowed, divorced, or obliged to contribute to the household income and commuted across the border for this purpose. Those who had crossed the border for a long time usually had their husband involved in small-scale trade during the heyday of the economic boom in the 1970s and, thus, were lucky enough to make enough savings to build a shoddy cement house in the town as well as in the village of origin or on the small vineyard land of the family.

Despite the escalating violence along the Syrian border, the townspeople held their firm convictions that the border gate would never be closed and small-scale trading across the border would keep up their livelihood.[1] A semi-legal trade is likely to constitute part of multiple work strategies needed to make a living at the border (Galemba, 2012, p. 10). This kind of trade was forged by neoliberal policies in Turkey as they introduced "the cross-border trade" regulations in several border cities, including Kilis in the mid-1990s. It was undertaken in the town as the primary or secondary activity by casual, precarious, or minimum wage male workers as well as women helping their family to make the ends meet, with little or no investment money for a small-scale venture as a sole trader or as a hired carrier. Non-governmental organization (NGO) and local bureaucracy representatives estimated that about 1,000 to 1,500 families made their living directly out of it.

The neoliberal policies revived the transnational shadow economy by providing the legal framework that extenuated the security measures and penalties, if not completely abolished, against the unregulated economic activities deemed illegal. But, the state's concern with the security of its territorial boundaries resulted in whimsical decisions and brought frequent changes in border surveillance as well as customs regulations. The town mayor stated in his interview that the arbitrary policies of the border acted as the barometer for the applications of the urban poor filed at the municipality for employment and social assistance. Any changing measure regarding customs regulation and border security could be translated into an increase or decrease in applicants depending on its effect on small-scale trading.

The town was rumored to harbor an army of unemployed and low-income dwellers who fell outside insurance coverage.[2] The social assistance and temporary employment schemes by the municipality and governorship, associated with clientelistic relations between town dwellers and authorities, were not sufficient to provide secure and permanent access to a living above poverty level, though they generated a response.[3] For example, the Social Assistance and Solidarity Foundation provided in-kind assistance regularly to 4,200 families and ran a soup kitchen where up to 325 families made use of it during the winter times. During my field study, I observed that the appointment decision to the Provincial Directorate of Central Employment Agency mattered a great deal among AKP followers in the town and became a point of contention, as each group of followers competed to exert their influence upon the new director to make him favor their own social networks for job placements.

The dependency on social assistance was also relevant in the border villages. The rural poor made their living by small-scale subsistence agriculture or casual and seasonal laboring that barely yielded revenues. The impact of globalization on agriculture led to the dismantling of supportive mechanisms for peasantry (Keyder and Yenal, 2011). The diminishing of price support schemes and the repealing of agricultural subsidies had an adverse impact on local agriculture by devaluing the price in agro-industrial products such as olive, pistachio, and grape in particular, which the peasantry relied on in dry farming. The decrease in the price of grape, for instance, was even more dramatic after the privatization of Tekel factories, the foremost buyers of grape cultivated by local producers.[4] The privatization of Tekel Suma factory in 2004 even frustrated the farmers engaged in large-scale commercial agriculture, as my interviewee complained that he had to rip off 10,000 grape seedlings after its transfer to the private sector. The town center continued to receive rural migrants in the last decade despite its lack of employment opportunities.[5] Thus, the rural poor turned to the contraband of cigarettes and livestock across the minefields whenever they could.

The interviews with the deputy governor and state institutions[6] put forth that cross-border trade construed a means of livelihood for the town dwellers. My interviewees' preference of these terms over 'smuggling' and the shift in the language away from emphasizing the illegal practices seemed purposeful as they probably assumed their position as state officers talking to a graduate student from Ankara and avoided to admit the violation of law under their jurisdiction. But the urban and rural poor strata assimilated the state notion of illegality by calling their practices of income generation "unlawful." Galemba (2012) argued in her study on the Guatemalan–Mexican border that the unequal access to such semi-legal trade is the key to understanding "the cultures of contraband" beyond the price differentials between two countries. Why were the poor dwellers of Kilis engaging in more risky and less lucrative ventures with the threat of confronting law enforcement or even death? Border scholars suggested that the enforcement of law and illegal practices of trade justified each other in the context of a balance between security and profit (Heyman and Campell, 2007; Galemba, 2013). The heightened border security would decrease the competition among the border dwellers and yield more profit. While the local government and broader border policies selectively enforced surveillance barriers and legal penalties against illegal crossings in Kilis town, they also provided space for border dwellers to resist, manipulate, or transgress them.

Profiteering from cross-border trade

When my interviewees from the Chamber spoke about the implementation of cross-border trade in Kilis, they expressed their frustration with the practice.

Though officially sanctioned and formally regulated, the trade underwent an abrupt cessation with an anti-corruption operation. My interviewees likened the operation to the military coup of September 12, 1980 in order to highlight its adverse effects. According to the former head of the Chamber, the operation made the border gate cease to be an international port. Öncüpınar gate of Kilis remained an A-grade crossing point, allowing the custom procedures for international import-export and transit trade as well as for goods declared by small-scale traders as personal belongings. What my interviewee probably implied was that the favorable conditions for large-scale entrepreneurs to trade on the legal loopholes had vanished. For example, the entrepreneurs from Gaziantep diverted away their direction from Kilis to Cilvegözü gate of Hatay or Habur gate of Şırnak to ship their goods or to make the transit trade to the Middle Eastern countries.[7]

The Parachute operation, true to its name, aimed for a secretly held raid by centrally appointed officers to a prominent businessman from Gaziantep with an industrial facility in Kilis as well as local bureaucrats in Kilis and Gaziantep.[8] The investigation involved the custom directors and governors in both cities, the customs director of Habur and undersecretaries of trade and agriculture ministries. The operation included a series of successive anti-corruption probes targeting upstart tycoons and high-placed bureaucrats. The probes received extensive coverage in national newspapers, indicating that the state budget suffered a loss of TL 3.5 quadrillion from unrighteously placed export-import subsidies to fraudulent trading. The businessman and his several company employees were accused of the fraudulent export of sugar, banana, tea, rice, and oil and the siphoning of TL 500 billion from the public treasury. The probe eventually grew to merge with other probes targeting local entrepreneurs in Kilis, accused mainly of the fraudulent export of sugar (NTV, 2001). The cases brought against them were eventually dismissed or timed out and the businessman from Gaziantep served a small amount of time.

The news detailed that the businessman did not only exploit the cross-border trade regulations, but also abused the inward processing and transit trade regimes (Baransu, 2000). The inward processing was introduced as a modification in the export incentive regime after the signing of the Customs Union between the European Union (EU) and Turkey in 1996.[9] With the inward processing, the exporters could benefit from tax exemption or reimbursement if they imported raw materials in order to process them and export the final goods. The regime subsidized the exporters in import duties and value added tax (VAT) during importation. Besides, the regime offered the exporters the subsidized prices for domestic raw materials if they guaranteed their processing and the exportation of final goods. On the other side, the transit trade regulations exempted the exporters from paying import-export taxes and customs duties. The businessman carried out export tax fraud by abusing these trade regimes and sold the goods on the domestic market although he pretended to export them. The authorizations and paperwork for fraudulent transactions involved the collaboration of state authorities and customs officers.

The businessman was also suspected of oil smuggling in the hiding places of truck trailers. The newspapers covered the allegations that the businessman ran a daily traffic of 150 trucks, suggesting an oil trade bringing in nearly TL 2.5 billion of unlawful profit (Hürriyet, 2000a).[10] The public prosecutor investigating the profiteering which occurred over a span of six years, commented that the investigated sugar business was less risky and more remunerative than the heroin trade (Hürriyet, 2000b).[11] The news coverage highlighted that the domestic sugar producers were implicated among countless anonymous reports that informed the state authorities against the ongoing profiteering at the border gates.[12]

The Parachute operation interrupted cross-border trade in Kilis but it did not extinguish it. The regulations of cross-border trade remained in effect, though amended in the late 2000s. So, I do not consider the anti-corruption discourse, rampant in the early 2000s, as a tool to fight malpractice and profiteering. This discourse served to bring such wealth generation practices under control in order to ensure fair competition in the market, as the news about the sugar or banana companies complaining about the profiteering at Öncüpınar customs implied. The anti-corruption discourse particularly took effect in the restructuring of the public sector after the financial crisis of 2001 and ensured the continuation of neoliberal policies, abating its destructive repercussions (Bedirhanoğlu, 2007, p. 1250).

For the border anthropologist Galemba, the unregulated trade across the Mexican–Guatemalan border constituted a neoliberal politics of invisibility, with the local authorities turning a blind eye to the illegal crossings at border zones that lost their significance as an international gates: the officers could not investigate the crossing trucks "unless the smugglers are not caught in the action" (Galemba, 2013, p. 7). This politics technically reduced the duty of Mexican officials to validate whether the truckers had authorization from the customs directorate at the region or capital city. In this vein, I suggest that the law enforcement at Kilis border appeared to function not to terminate the unregulated trade but to take it under control, although several legal reforms were undertaken to prevent smuggling and corruption under the auspices of neoliberal governments after the 2001 crisis.

Livelihood expectations on the eve of the Syrian conflict

The semi-legal trade that evolved after the introduction of cross-border trade regulations differed from the transnational shadow economy of the import-substituting industrialization (ISI) period as the lower strata had to undertake the risk, though their labor could still be exploited by the large-scale entrepreneurs. In the absence of patronage, the lower strata faced several hardships, including downswings, bankruptcy, and imprisonment. While the neoliberal policies blurred the boundaries between the legal and illegal realm by regulating "cross-border trade," they ensured its unequal access by various strata. The regulations since the mid-1990s not only took unregulated trade under control by turning a blind eye to the lower strata's struggle to earn their living, but also sustained an income differentiation among them. The unequal access to trade constantly fed the prospects of upward mobility for the lower strata, although the few, not necessarily from rags-to-riches, could prosper. The lower strata recognized the unequal access as a discriminatory practice, but tried to fight it by improving their chances on the basis of their social networks.

The low-income dwellers of the town recalled the initial practices of cross-border trade as the "sugar period." The sugar business not only helped local entrepreneurs prosper, but also small-scale traders commuting back and forth every day in order to collect sugar from the opposite Syrian border town Azaz. My interviewees[13] asserted that sugar was unloaded from each car crossing back over the border. They did not have particular knowledge about the modifications in the legal framework that initiated or ended cross-border trade. But they were conscious of the policy that encouraged the custom officers to disregard the overcharging of car trunks with undeclared Syrian goods. According to a young commuter, "the smuggling was set free" in that period:

> This government enforced the Customs Laws as soon as they came into power. You probably know that. Now, the legal right of any person is three kilos of tea and five kilos of sugar because the Customs Law is made. The cigarettes are banned. That is to say, before the Customs Law is made, the border crossings were set free. I unloaded the car myself. We used to take out a ton of sugar from the car.[14]

Besides the trunk, every hole under the hood could be stuffed with sugar sacks. My interviewees recorded that a kilo yielded one lira of net gain. Thus, a commuter could earn TL 1,000 in a trip.[15] The gainful trade raised the prospects of upward mobility for a larger population in the town, even attracting small business owners and employees for additional income.

The sugar period, as my interviewees called it, reintroduced unregulated trade as a mundane effort of earning a living for broader low-income strata. Even though they had to sell their goods to local large-scale entrepreneurs, collecting the sugar in their high-walled garages near the beltway to the border gate and shipping it to the domestic market in trucks, they were not dependent on patronage relationships to find employment. Still, the lower strata had to rely on political patronage in order to ease the customs control at the border gate. According to Serdar, a subcontract employee at the public hospital and a small-scale trader, the deputies could pull votes from a large constituency when they put pressure on the customs officers to turn a blind eye to the border crossers with undeclared goods:

> It [the border control at the gate] was so tight for a couple of days just after the elections. I saw it with my own eyes. A deputy came to the customs area. He asked to the customs director, "[do you realize] why the military posts did not work". I was there. [The customs director said], "they bring so much sugar". The deputy raised his hand and slapped in the face of customs director. I saw it with my own eyes. He [the deputy] wandered around and came back, "do your duty right or you have got no business here". He slapped him in public and sent back. Then the gate was opened and the crossings was set free. No inspection! I mean, a deputy can make the gate work if he wills.[16]

The loss of profits from the decline of the sugar trade was swiftly replaced by another remunerative business: "Now, the sugar is over and the cigarettes business started. Not anybody could bring three cartons of cigarettes at that time. Today one can bring fifty cartons."[17] The town dwellers kept commuting across the border in order to bring Ceylon tea, cigarettes, mobiles, automobile spare parts, home hardware, and cheap quality, showy goods. The oil trade consisted of selling the extra oil in the car's tank. The regular gas tanks were changed to bigger ones until the legal amendments occasionally banned them. These cars with higher road clearance allowed the carrying of up to 100 liters of oil, doubling the size of oil that regular tanks could contain. Female peddlers were helped by their dresses and topcoats in order to brace cartons of cigarettes with laces on their legs and arms. They also tended to buy and resell women's underwear, cotton domestic cloths, cosmetics, and perfumes of poor quality.[18] The hiding places of cars could be used to not only carry these goods, but to also sneak parakeets out of sight, which yielded more profit than the catchpennies. The lifting of visa requirements between the two countries in 2009 further endorsed cross-border trade as a rampant opportunity for earning money. With cross-border trade, Kilis developed a symbiotic relation with the Syrian border town Azaz, located 7 km away.

Similar to Kilis, Azaz was economically dependent on unregulated trade and the town dwellers were involved not only in selling goods to the Turkish traders, but they also provided out-of-sight places for buyers to store their goods in the hidden slots inside their car. The garagists, as they were called, let the drivers use the high-walled rustic courtyards as garages to store their goods. Usually, the garagists fetched the order of traders from outside for a small commission. Storing goods in the car in sight could put the Turkish traders in jeopardy of getting imprisoned in Syria. The garages were not only places where the drivers could store their purchases but they could also stay overnight in their car when they had to wait for the shift of the customs officer that their garagists had contact with or for the ending of a high-ranking officer's spot check at customs. Thus, the initiation of cross-border trade narrowed the range of the transnational economy down to the distance of 7 km, which the dwellers commuted as much as the regulations allowed.

The old dated anti-smuggling law was amended and replaced first by a liberal penal doctrine in 2003 within the context of EU accession (Bozçalı, 2014). The 2003 law enforced cash fines for smuggling and the traders could avoid jail time. When a new law was legislated in 2007, it restored the penalty of imprisonment as well as the confiscation of vehicles used in smuggling.[19] Accordingly, the traders faced jail time of between one to five years for smuggling goods across customs, the imprisonment being severe if the goods were smuggled across the border zone. If committed for oil smuggling, the trader would face the penalty with a minimum level of two years. Despite these penalties, however, the local dwellers continued to commute across the border with undeclared goods in exceeded quotas.

The quotas sanctioned by the regulations of cross-border trade for the goods that a passenger could bring were so low to allow small-scale traders to yield a profit. For example, a passenger could only carry a kilogram of tea and 400 cigarettes (two cartons of cigarette packs) through customs as personal belongings. My interviewees stated the net gain for a kilo of tea as TL 2 and for a cigarette carton as TL 2.2.[20] According to them, the allowed quota was not even enough to cover expenses, including the departure fee of TL 15. Thus, the trade would only render a profit if the trader abused the allowed quota. Ignoring the officially sanctioned quotas, the traders groped for a limit at the optimum reached as the result of long-lasting disputes at customs, the traders arguing for their rightfulness to pass with undeclared goods in order to earn a living. The words of Talip, a young small-scale trader, reflected this viewpoint: "The guy brings five or ten cartons of cigarettes. They take it away from him. He brings home the bread out of it. Is this not a pity?"[21] In the absence of patronage, their business was jeopardized by the risk of having goods confiscated. They could not afford to bribe the officers, so they had to count on their network, taking advantage of the confusion due to the crowd and disarray, trying to intimidate or make them pity them for not seizing the smuggled goods in their cars, bags, or beneath their cloths. My interviewees told of several cases in which the customs officers were mobbed by a furious crowd and the managing officer was manhandled.

An informal conversation at the border customs would also confirm it. The customs building, located at the entry of no-man's land, appeared neglected with no signs directing the visitors inside. I was not able to find the director's office with whom I had an appointment unless the secretary guided me within the building. I was not even sure about the title of the person to whom I paid a visit since I was only given a name on my prior phone call. As I introduced myself, he told me that it was not possible for him to give me information since he was a public officer. Yet, he assured me that he could answer my questions informally. He was called chief by the officers. So I assume that he was the chief of the Smuggling and Information Office at customs since he mentioned this was his previous duty in the late 1980s before he had was appointed to Kilis. The anonymity within the building was intentional. The highly-placed officers had to be invisible in order to escape the fury of traders. Most of the officers used to live in Gaziantep rather than renting a flat in the town for security reasons. The chief complained that the town dwellers saw the border gate as their own courtyard and they did not acknowledge state authority. When he tried to enforce the regulations, he could run up against a townsman pulling his knife. According to him, the dwellers had the habit of crossing the border first thing in the morning as if they were going to the coffeehouse after they woke up.

Nevertheless, when caught out by the customs officers, the traders were not reported to the police. The customs officers, rather than taking criminal action against the traders, held the goods and moved them to the customs warehouse. The traders could then claim their goods from the warehouse at a particular day during the week, but they had to cross first to the Syrian side with the retrieved goods and they would try to enter anew. When they retrieved the goods, they also had to take the risk of getting them seized either by the Syrian or Turkish customs. Recently, the traders had to pay an additional warehouse rent of TL 50, which worked practically as a penalty for the traders, leading them to leave their goods without ever reclaiming them. On the other hand, the traders always came up against the danger of coming to the officers' attention. When caught out, their cars could be towed away, followed by a lawsuit brought against the owner. But, the interviews implied that the customs officers tended to overlook small amounts, a tendency indicating to the legitimacy of small-scale trade as a means of livelihood. According to Bahtiyar, a young worker, the cars could come to the attention of officers, but they used to let the peddlers pass. But in the absence of patronage relations, the seizing of their goods or the fines imposed exacerbated the economic woes and growing debts of lower strata, especially for initiators in carrying goods in exchange for a hire.

The young man had left his job as a molder in a workshop, where he was paid a daily amount of TL 20 since his salary "did not save the day," in order to smuggle goods across the border for a trader. I had run into him in a stationery shop that I paid a visit to most of the time I was passing by. While the shop was a place to drop by for having tea and a nice chat with the owner, it was haunted by the poor urban dwellers having their papers Xeroxed and getting help from the stationer with filling in the application documents for a Green Card, a free health benefit scheme provided to the poor not covered by public health insurance. The shop was right across the building where the Green Card office was accommodated and the stationer was a caring man, not declining any requests for help. It was easy to tell at a glance that the worker and his young wife standing next to him with a little baby in her arms appeared poor. As the stationer knew about my research, he made them sit in order to let me ask my questions in haste. The young man had just started to bring parakeets and aquarium fish across the border for a trader in exchange for TL 50. More remunerative goods like tea and cigarettes could be carried by experienced carriers for TL 150.

He used to cross the border by making a car stop to pick him up, since pedestrians were not allowed to walk across the no-man's land between the two gates. He went to Azaz town to collect the order and then came back across the minefield near the gate towards morning. He could carry two big nylon bags at a time, each containing a thousand small aquarium fish in smaller bags. As soon as he crossed the mined land, a car arranged by the trader would pick him up. To him, the officers let the peddlers cross into Kilis through the gate because they regarded them as "wretches" living in dire straits. But he had to turn back the other way around, across the minefield in order to escape the cameras at the gate. The peddlers were only porters. Since they did not own the goods, the loss would be covered by the trader if the peddler was caught out. But without the patronage, they or their family had to stand the financial difficulties when served with a fine or jail sentence. After a short while, I dropped by the stationery shop and learned that the young man had been caught in the act. The stationer told me that he did not have to pay for the loss of the goods with him, but he had been sentenced to a substantial pecuniary fine for violating the passport law, which he could not afford to pay by any means. The young man came to the stationer in desperation, and he helped him write a petition to the governorship asking to be pardoned.

The peddlers could not afford to bribe the officers, so they had to take advantage of the confusion due to the crowd and disarray, trying to intimidate or make the officers pity them for not confiscating the smuggled goods in their cars, bags, or beneath their clothes. Left to their own means, they could only engage in bringing goods in small-scale. Although the customs officers tended to turn a blind eye to them, the crossing of undeclared goods was not without difficulty. The way that the customs officers treated the lower strata border-crossers during the inspection offended them. The plastic bags of the peddlers were cut by the officers to enable better inspection. The officers could ask them for a body search or to take their coats off. The way they treated the peddlers was likely to offend them and reveal the unequal access to cross-border trade in the viewpoint of the lower strata, as my interview with a woman peddler highlighted. In her seventies, she was still commuting to Azaz for small-scale trade at infrequent intervals though she had moved to Gaziantep city a long time ago. She had no pension left from her late husband and she maintained her life in a one-room basement flat thanks to her brother who owned the two-store building.

I remember that, sat on a pillow on the cement floored ground, I was having a chat with her when she was sitting on the bed with her legs extended. My visit was to check on her since I heard that she had severe back pain. She spoke about her fears of having possible surgery for the hernia in the middle of her back, since the surgery would impair her movements and put her back from peddling. Then she started to talk about the unfair treatment she was receiving at the customs gate. She lost her 10 kilos of tea to the officers last time she was returning from Azaz. She said she was offended that the customs officers cut her tea packs while they let the trucks with smuggled goods pass. The conviction that the large-scale entrepreneurs could have their trucks pass unsearched by the border customs prevailed among the lower strata. Nesim, a border peasant engaged in smuggling cigarettes across the border, expressed it in a rather racy manner: "The gate is a camel, an elephant. The company blatantly makes the biggest haul. But the officer catches the poor wretches like us and he happens to do his duty."[22]

The lower strata commuters were aware that the transport companies unequally traded on the profits of border trade as they could afford bribing the customs and covering the damages of getting caught. The small-scale traders regarded the amendments in the legal framework of cross-border trade as arbitrary changes in the border policies which could be enforced both by the government or local authorities. In the words of Talip:

> I suppose, it was three years ago. [Crossing to] Syria was allowed once every three days in 2008. Afterwards, the daily entrance and exit by car was set free in the last two years. But after the parliamentary election [in 2011], it was regulated back as once a week. You can cross by car once a week and everyday on foot.[23]

My interviewee Talip did not know that the last restriction on the frequency that the personal and company vehicles might go abroad was not a formal ban. A decree enforced by the cabinet in the early days of 2011 aimed at stopping oil trade smuggled in the vehicle tanks by imposing a fixed tax of TL 150 for the unconsumed oil if they went abroad more than once in a week and four times in a month.[24] For the small-scale traders as well as owners of passenger transportation companies, this practically meant a strict limitation on their movements and an obstacle to their engagement in trade. Talip explained it as follows:

> For example, we used to commute everyday. About 100 to 150 liras was left daily. Now you drive and with a bit of oil, you can only yield up to 300 to 400 liras. If you gain 150 liras everyday, it makes 750 liras a week. When it was everyday, opportunities were better. You brought and sold, and you went back again the next day. We had an occupation after all. Now we do not have it either. You go once a week and you are idle.[25]

Thus, the small-scale traders found it convenient to organize among themselves to form informal 'companies' to overcome such restrictions. They collaborated with their relatives or acquaintances and, as profit-sharing partners, they shifted their cars so as to get around paying the tax.

The small-scale traders tended to run their business by taking on debt, especially when they were engaged in the retail trade of cigarettes. Thus, the small-scale trade ran on the basis of the debt economy not only in Kilis but also in other parts of the southeastern border of Turkey. In Yüksekova, for instance, the rising penalties and growing risk of getting caught recently led the cigarette traders not to use carriers despite the profit margins in unregulated trade from Northern Iraq having increased (Özcan, 2014). Thus the carriers had to pay for the goods themselves and took the risk for any possible losses. The wholesale trade of cigarettes in Kilis consisted of its trafficking across the minefields and its sale to the local shops or its transportation to Gaziantep where local shop owners bought the goods. The border villagers needed to undertake security and financial risk: they owned the goods on credit, crossed the minefield at night in order to pick up the goods from Syrian peasants at the border, left at a predetermined point at the border and transported them at daylight by using side roads to Kilis or Gaziantep. If they were caught out, then, unable to pay off the debt, they would have become more dependent on unregulated trade to pay it by leaving another unpaid.

The arbitrary border policies and absence of patronage to cover their financial losses forced them to count on their own networks to survive the rise and fall of trade. The town dwellers were usually involved in unregulated trade because of its intergeneration transmission as an occupation. In some cases, they could draw on their parents' networks and take advantage of these relationships to have smoother access to unregulated trade. Hamit, for instance, benefited from his father's connections at the border customs. His father is a loved *imam* in the border villages where he used to work, thanks to his integrity not only as *imam* but also as smuggler. He presently served as *imam* in a mosque near the customs, remained at the former rest area accommodating prospective pilgrims going to Mecca by road. This mosque was presently frequented by the customs officers during prayer time and the *imam's* good relations with them helped Hamit's business: the departure and custom declaration procedure was made easier for him.

Hamit and his brother run a small grocery shop where they also received potential customers looking for cheap oil. Although the Syrian conflict caused the rise of oil prices, it was still profit-bringing for the commuters due to the price differentials. For instance, the oil was "cheaper than water" in Syria, the price being one lira per liter.[26] The grocer in my neighborhood said that he could easily sell the oil for TL 3.5 per liter to his potential customers, with an increase in price due to the added up profit of the commuters carrying the oil for him. Hamit and his brother commuted to Azaz alternately, one or the other left to look after the shop, to bring cigarettes as well. He had started to commute across the border since the age of 17, to carry oil in bins on a hand truck and he had made money since then to own a house and cars, including a modified one with a bigger gas tank. It was the time when the violent clashes in Syria reached out to Azaz and the town was being bombed by the Syrian fighter jets for the first time that I wondered how much longer he could pursue the trade across border. He was quite satisfied with the state of business and expressed his conviction that the border gate would not be closed despite the increasing tide of the conflicts. He went into partnership with a fellow townsman and rented a cabstand place from the municipality. Planning soon to open it, he explained that he could presently afford individually to run three taxis.

The fact that Hamit's business went well illustrated how the unequal access to trade could differentiate the prospects of living a better life for the lower strata. My interviewees thought that the prospects for upward mobility depended on the courage and luck that anyone engaged in "smuggling" could have. Not necessarily from rags-to-riches, but unregulated trade could provide a middle-class lifestyle for the families with a lower strata background. Serdar living next door in Kilis was exemplary. He had commuted to Azaz since his youth. He lived on a housing estate, first built by the municipality to provide accessible housing for the lower strata, but then turned into a project that the middle-class profited from with the favorable terms of sale. The housing accommodated families mostly with male heads working in the public sector or employed in regular jobs with social security. Teachers and police officers were outstanding among the housing population.

Serdar and Arzu owned a flat that was furnished in accordance with middle-class taste: a large guest room equipped with an elegant sofa set with wooden engraved frames, a buffet, and low tables. The couple owned a foreign brand car that Serdar used for carrying oil, cigarettes, and other tradable goods. The couple could afford to go to the private clinic in the town every time their little boy was sick. As I was conscious of her husband beating her occasionally, I realized that her marriage was regarded by her parents as a good chance for her to maintain better living standards than her family could provide. Arzu's mother used to come to my flatmate for house cleaning once in a while and she seemed to enjoy fulfilling her duty as a good mother to find a good chance of marriage for all her daughters. Serdar was a young man who could establish himself as the head of a family with soon-to-be five members. Arzu fell pregnant with her fourth child—one being deceased when he was too little—at the time I was living there. Until he found a blue-collar job as a desk worker provided by a subcontracting company in the state hospital, he had survived tough times and had been in both Turkish and Syrian jails for short times.

For him, finding a job with social security was probably more decisive in establishing a more stable course of action in small-scale trade. Serdar continued to do small-scale trading while he worked at the hospital. His night shifts allowed him to commute to Azaz in the rest of the day. Although he could cross the border once a week in order not to pay the tax, he was able to earn money more than the minimum wage that he received from his hospital job. While his wage was near TL 650, trade yielded about TL 1,000 monthly. A secure job permitted him to assess the risks imperturbably and retain a low profile, avoiding drawing the attention of customs officers as much as possible. His interview reveals that the Turkish customs sought to take the small-scale trade under control, allowing the lower strata to earn money while keeping the trade definitively small-scale:

> If you lose a couple of cigarette [packs], it is not a big deal. But if I go by car and my 5 million worth of goods are all seized and they sue you, then I cannot compensate it in a year. There is a difference between losing the goods and losing the goods. The guy [customs officer] came and took some of my tea and let me go. That is nothing. He checked my car and he did not take anything. People bring parakeet for a hire of 100 liras. You take it and bring here and you get 100 liras. The inspector heard the bird call as he passed by. He looked at my face. I did not say anything. He had checked the car. He came back. He put the notebook on the floor. He leaned down and looked and he noticed the bird from below. As he saw the bird, he told me to open the trunk. I opened the trunk. He took my five kilos of tea away and threw it to the warehouse. If he had not seen the bird, he would not have taken away anything.[27]

Serdar also gave his opinions about the nature of unregulated trade:

> You have the crank or the carburetor. He says, for example, "Take it [an automobile spare part] and you will have fifty or hundred million [liras]. Not all men would take it. It is risky. If the inspector notices it, he will not let it go. It is a profitable business but it is risky. You need to have such a place in your car to carry it. For example, he saw the bird in mine and he took five liras worth of tea. That depends. Bit of luck and vigilance. Not everyone is tailored for it. It is not everyone's business.[28]

As the above excerpt indicates, unregulated trade was encouraged by both Turkish and Syrian traders and entangled the traders, commuters, and state officials of both sides in a symbiotic economy. As it is suggested in this section, the cross-border trade regulations enforced by the Turkish governments unilaterally constituted a phase of neoliberal politics, promoting profiteering through fraudulent exports and turning a blind eye to unregulated small-scale trade as a means of livelihood for the lower strata border communities. Even though the regulations of cross-border trade required a complex procedure of application and paperwork, the small-scale traders in Kilis could engage it without necessarily going through it. Small-scale trading ran on the basis of the debt economy and jeopardized the traders with risk and uncertainty, promising upward mobility to the lower strata but granting only differential access.

Giving bribes, taking brides: the politics and economy of kinship among the poor

As mentioned in the preceding chapter, the arrival of Syrian migrants triggered the discontent among town dwellers and overshadowed the cross-border kinship ties that they laid much emphasis on. But, the interviews were likely to highlight that the open-border policy with Syria increased encounters between the two communities since the mid-1990s. They revealed the border as a space of flows where goods in return for bribes, greetings, and brides were exchanged. As a scholar of the Central Asian porous borders suggested, "encounters with 'the law' are frequently negotiated through appeals to common ethnicity or religion, or trumped through the law-dissolving agency of the bribe" (Reeves, 2007, p. 73). Kinship and marriage bonds both normalized and strengthened the exchange practices across the border deemed illegal by the authorities.

I address the shift of kinship to a material strategy and discursive tool for the lower-strata town dwellers in navigating legal as well as social boundaries. While the interviewees' recollections about the import-substituting period revealed the significance of kinship relations in facilitating the cross-border exchange, the conversations about kinship in the open-border policy period laid emphasis on the social boundaries between Turkish and Syrian societies. According to the interviewees, the open-border policy between two countries had increased encounters and forged a common economy. However, curiously enough, they also tended to emphasize the backwardness of Syrian society and reinforced the 'Arab perception' of the local community.

Before the news about the Syrian migrants was communicated, my conversations with the town dwellers highlighted the cross-border kin and marriage relations, presumably heightened by the recent open-border policy of the Turkish government. First, it was puzzling for me to observe the shift in local speeches about the Syrians from kin to unwanted guests. But in a post-research visit to the town for a week of translation work in an international NGO in October 2012, I had the chance to witness a shift of perceptions among the locals that had kinship relations with the coming migrants or who had contact with them.

Kilis proved to be one of the border towns where the Syrians depended on their kinship ties in establishing themselves as migrants living outside the camp (Özden, 2013, p. 3). The Syrians had already started to work in local cafes and eating-houses and in the fields as agricultural laborers. The local marketplace was enlivened thanks to the shopping needs of Syrian families. The visit to the town also gave me the chance to pay a visit to the salesman of the second hand store whom I used to frequently visit and chat to during the field study. He was busier than ever and received Syrian customers who came to the depot-like shop to pick up, for instance, needed furniture for their home or an old television to cheer their prefabricated house in the campsite. Despite our conversation being interrupted by the arrival of potential customers, I kept inquiring from him about the reactions regarding the overgrowing presence of Syrian migrants. He solved the puzzle for me as he said: "Do not pay too much attention [to what they say about the Syrians]! We are kin in the end".[29] The recurrent trope of "cross-border kin communities" did not signify a turnabout among the local community about the Syrian migrants. The discontentment and cultural tensions were reported to continue (Dinçer et al., 2013). Nevertheless, it implied the need for a more sophisticated understanding of kinship relations across the border.

Cross-border exchanges, often expedited by the ethnic or kin relations, may reinforce social boundaries rather than undermining them. But few studies thoroughly examined the changes in kinship relations catalyzed by the re-bordering processes. Parizot's study on the cross-border ties among Palestinian Bedouins remaining in Israeli-occupied Negev and self-governing Palestinian enclaves indicated that the kinship relations could act as an economic favor rather than social belonging "in a highly unstable political and spatial context, where borders are constantly redrawn and the statuses of people are redefined in ways that have often broadened the gap between cross-border 'partners'" (Parizot, 2008, p. 70). Parizot wrote that the smugglers, for instance, needed tighter coordination for their activities between divided territories. As the Israeli state increased security measures after the Oslo process of 1994, the legal or illegal crossings to Israel across the separation line required the invitation or networking of Bedouin kin, raising new trans-border configurations of power and heightening the perceptions of difference and even social distance among the kin.

I discussed earlier the persistence of paternalist relations within the context of the transnational shadow economy. In the same vein, I argue here that that the growth of shadow markets helped the reproduction of kinship relations rather than their decline. The field observation and interviews hinted at the ways in which the mining of the border and the enhancement of contraband in value affected these relations. As mentioned earlier, the extended families in Kilis could be derived from the chiefdoms or tribal confederacies which were actually territorially-based socio-economic organizations, rather than simple lineages. Though the administrative centralization renders them smaller and more heterogeneous, their engagement in contraband and tendency to establish family monopolies around their illegal businesses resulted in the persistence of tribal ideology[30] as a social and economic unit.

Before its mining, the border could be crossed by walking through the stones barely marking the frontier line, especially in the mountainous Kurdish region where the military patrols could be eluded more easily in comparison to the lowland villages. Kurdish peasants used to slink across the border to the Bilbilê (also Bulbul in Arabic) town of Afrin in order to exchange their eggs and crops with goods like medicine and soap. The contraband trade rapidly evolved in the 1960s to make the poor border dwellers thrive beyond the dire straits they used to live in. A former Kurdish smuggler who was now running a grocery shop in the town expressed that he bought his first car at the age of eighteen and no fellow villager had ever seen so much money. His adolescence was full of memories about border crossings on foot in order to stay at the Syrian village of their relatives for weeks, work in the fields driving their tractors in exchange for hospitality. As the words of Müşir, a former smuggler, illustrate, after the consolidation of the border with fences and landmines, the kinship relations shifted into business connections and the exchange practices among kin were shaped by the money economy, though trust remained an important component:

> The kin bring the goods but on commission. We brought everything we could. The traders waited for us in Kilis and usually [they came] from the west. One can also work jointly with a kin. If you do not have money, the kin gives you the capital and simply gets a commission. But only the kin would do that.[31]

The Turkish–Syrian borderlands attested to the significance of kinship and affinity for business relations across the border, not necessarily restricted with the unregulated trade. In Hatay, for example, the ethnic communities across the border utilized their "bonding" and "bridging" capital to promote their business relations beyond the constraints of ethnicity and religion (Doğruel and Karakoç, 2013, p. 8). In Kilis, new alliances among kin established by cousin marriages helped to maintain the trade as family business. As I observed among my interviewees, not only did the Turkish traders cross to the Syrian side, but the Syrians also came to the town in order to sell the goods that they had brought in with them. For the lower strata of both sides, the relatives across the border constituted a social network to facilitate border crossings, providing grounds for arrival, lodging, and contacts for making exchanges.

Though the kinship bonds were weakened by the regressed trade across the border after the military coup, the town dwellers believed that they were heightened with the open-border policy of Turkey. Turkish governments mended the fences with Syria in the late 1990s with the ceasing of the water dispute and the ending of Syrian support for PKK, the Kurdish Marxist–Leninist insurgency.[32] Nevertheless, the opening of the geographical border may not automatically contribute to the opening of social boundaries. While it encouraged local discourses on pluralistic openness and the multiculturalism of Hatay (Doğruel and Karakoç, 2013), it tended to magnify the perceptions of difference among border communities in Kilis. As a conversation with a local white goods distributor married to his Syrian cousin revealed, the corruption of the Syrian bureaucracy and concomitant obstacles to cross-border trade were interpreted as a major concern in recent times.

The protocols and agreements signed between Turkish and Syrian authorities were the result of two countries converging on political and economic cooperation since the late 1990s. The Free Trade Agreement, signed in 2004 and ratified in 2007, was followed by a number of partnership agreements in the fields of energy, tourism, military training, study missions, technology exchange, and security. Apparently, the convergence between the two countries proved fruitful in heightening the border crossings as well as cross-border kinship relations. During the conversations with town dwellers on border life, the Syrian brides stood out as a popular topic that they wanted to put emphasis on. In 1999, the Kilis governor heralded the signing of a protocol with the Aleppo governorship in facilitating the exchange of greetings in religious holidays by sanctioning the temporary residence of families at the other side of the border (Yeni Şafak, 1999). The protocol was enacted a few years later and, since then, the Turkish and Syrian families took turns each holiday to cross the border by showing their identity card only and staying for two to three days in their relatives' home.[33] In 2009, two countries signed a visa exemption agreement, which further promoted the border crossings. Thanks to these regulations, the families could avoid the tumultuous and chaotic meetings happening through the wire fences and benefited the loose customs control in engaging small-scale trade, crossing the border gate on their way back with undeclared goods. But the contribution of these protocols was principally the heightening of kinship relations and the increasing number of Syrian brides according to the interviewees.

The conversations underlined the complaisance of Syrian brides in settling their lives in Kilis town, wedded to a Turkish man and the 'backwardness' of Syrian society. Also, they pointed out that the practice of polygamous marriage was still legally in effect in Syria, as four places for spouses were allocated for the husband on the marriage certificate. Accordingly, the young Syrian women would readily accept to leave their native town since they would be emancipated from the familial and state oppression against women and as entitled Turkish citizens after three years of official marriage, they would eventually gain freedom that they did not ever have in Syria. These emphases were complemented by the comments on the corrupt business environment in Syria. Among the town dwellers engaged in cross-border trade, the belief that the Syrian customs officers encroached to reap personal profit from the "duties" paid by the Turkish crossers was prevalent. They suggested that the duties were going in the officers' pocket rather than to the state revenues in Syria as the officers compensated for their poor salary and social protection with extra profits.

The interviewees said that each car was charged about 1,500 to 2,000 Syrian pounds[34] or 65 to 70 Turkish Liras for their crossing regardless of the load they had with them. Also, they complained that they had to bribe the officers for any formal paperwork required for entering the country such as issuing a triptyque (a customs permit for temporary transportation of a motor vehicle) and travel insurance. The company owners protested the arbitrary practices of Syrian customs and the border control that lacked any counterpart on the Turkish side. As I learned from my interviewees, the Turkish and Syrian local authorities tended to enact regulations and practices regarding the customs and border protection in line with the principle of equal treatment in return. That is to say, if the other party imposed a favorable or unfavorable practice regarding the opposite country's citizens, the latter was expected to put the same practice into effect. The authority of Kilis and Aleppo governorships over the border gates allowed them to coordinate these practices. But there were many exceptions that the Turkish company owners groused about such as the overtime pay that the drivers had to give to the Syrian officers on Fridays and Saturdays, declared as the official weekend. They argued that Turkish drivers had to pay a couple of times more than the drivers entering the country across the Syrian–Jordanian border and the Turkish side entirely lacked it.

These complaints highlighted the complex organization of the border economy that prevailed in the twin Syrian town of Azaz. Moreover, the Turkish traders had to rely on the garagists, who brokered the profit-sharing arrangements with the Syrian officers. They intervened as the traders' "*alaqat*" (in Arabic, connection, intermediary) when they had any problems with the Syrian police or customs officers, especially in order to avoid the fearsome threat of prison. An oversized gas tank, for instance, could be an occasion for the Syrian police to levy a heavy fine. Then, the garagists mediated and reached a solution at the end of long and tedious bargains about the amount of bribe. The garagists in such cases could assume responsibility and cover full or equal share of the demanded bribe.

For Kilis locals, these practices, combined with the declining profit margins in cross-border trade culminated in the perception of the Syrian state as corrupt and, thus, backward. For Turkish traders, the border economy at the Syrian side was in sharp contrast with the Turkish side where a deterrent and punitive policy awaited them. Even though the better-off traders breached this policy by bribing the Turkish customs officers, they assumed the Syrian side to be corrupt rather than the Turkish officers and ignored the fact that the border economy straddled both. The effective enforcement of border control on the Turkish side was particularly effective after the anti-corruption operations against border customs and local authorities. Border studies rightfully pointed out that state intervention to combat illegal practices does not function to stop the transgressors to engage them, but "enables the state to selectively instill fear and assert sovereignty" (Galemba, 2013, p. 280). It also serves to sustain the status quo among different power groups, while vindicating the upper layers of state bureaucracy. As Akhil Gupta's study on the discursive construction of state in public culture demonstrated, the discourse of corruption circulated by the national and local media may inform symbolic representations of the state, reflecting the expectations of accountability both from the state officials and local groups. The local bureaucracy and discourses of corruption together construct the state as an "imagined translocal institution with its localized embodiments" (Gupta, 1995, p. 389). Similar to the convictions of Indian villagers that informed Gupta's study, the public culture in Kilis town seemed to distinguish between multiple layers and distinct locales and centers that constituted the state and confined the corrupt practices with the local authorities in contrast with the dominant perception of the corruption in the Syrian regime.

This did not exclude the possibility that the Syrian government could be engaged with transnational institutions and agendas promoting policy reforms towards transparency and accountability, particularly in effect during the neoliberal era. But Kilis locals were certainly not in the scope of such discourses. Hence, the encounters of Turkish traders with the Syrian officers and locals heightened their perceptions of difference that the traders construed it not only as characteristic of the Syrian state, but also Syrian society. Despite the emphasis on kinship relations, these encounters highlighted in the view of Turkish traders the cultural gap between the two societies.

I am reminded here about the negative stereotypes and images of Arab identity historically produced and reinforced by mutual nationalisms. The scholarly resources pointed to the emergence of these negative stereotypes and images incorporated into the production of nationalist discourses in order to raise 'nationalist consciousness' (Watenpaugh, 2005).[35] The political and cultural influences informed the Turkish images of Arabs distortedly depicted as Bedouin, traitor, womanizer, uneducated, unreliable, undemocratic, and submissive (Watenpaugh, 2005, p. 53). Hence, it was no surprise that these images were reworked to draw social boundaries between the two communities, when the news about the coming of Syrian migrants circulated among Kilis locals. In those days, several interviewees repeated the rumors that they heard about the unreliability and lack of honor of Syrian migrants in abandoning their homeland in wartime, a tendency to misbehavior outside the camps, and lewd demeanor of migrant women debauching the gendarme.

These social boundaries could even constitute battle lines within a household as in the case of discord between Asiye and her daughter-in-law. Her daughter-in-law, a Syrian woman relatively elderly for a first marriage, was the second spouse of Asiye's son. For Asiye who had been engaged in small-scale trade for 40 years, stepping out beyond the Turkish customs in order to cross to Syria meant to step on the Arab side, which clearly indicated her distance to them: "[After the passport check by the Turkish customs] we take our passport and we go and enter along the Arabs' [territory].[36] Her son, an outsourced worker in a high school fire room, was abandoned by his first spouse and the mother of his children as she ran off with another man. Thus, Asiye had to arrange her son's second marriage by using her social network in Aleppo to find a Syrian woman whose bride price would be more affordable compared to her Turkish counterparts. She seemed to never get along with her daughter-in-law although she had arranged the marriage herself. Her regret and discord would make her daughter-in-law almost automatically Arab in her eyes, though she came from an ethnically mixed familial background with a Turcoman mother and Kurdish father.

Certainly, even though the open-border policy and increased interaction might reinforce the social boundaries among border communities with longstanding trade and kinship ties, they still should be explored by addressing the entrenched hostilities between Arab and Turkish nationalisms and the working of nationalist ideologies at the community level. Nevertheless, as the discussion of this section aimed to indicate, the cross-border marriages and kinship relations proved to be resilient and versatile ways to maintain alliance between families at either side of the border. They normalized the border crossings, did efficient work of networking among the lower strata for sustaining the exchange practices, and helped the reproduction of kinship as a social and economic institution. In the light of this discussion, one can anticipate that new alliances among border communities built under highly unstable circumstances might alleviate the cultural tensions created by refugee arrivals in large numbers, though at the cost of further isolation and estrangement among the dwellers deprived of such allying ties.

Concluding remarks

The lack of patronage exposed the poor to the dangers of criminalization and penalization, both prison sentences and penalty fines, and getting into debt. The cross-border trade regulations introduced in the decade of the 1990s contributed to the individualization of coping strategies among the poor and undermined the potentials of cross-border trade to guarantee the collective right of livelihood by opening room for semi-legal but small-scale trade activities. Thus, with the neo-liberal policies, the shadow economy not only promoted unequal access among various strata, but also resulted in income differentiation among the poor.

Yet, this chapter also pointed to the cross-border kinship and alliance between families which proved to be resilient and versatile ways of sustaining illegal exchange practices. Although the open-border policy of the Turkish government facilitated border commuting and forged a symbiotic economy among the border dwellers at both sides, it reinforced the perception of difference and social distance. The poor, however, employed kinship relationships and cross-border marriage as a material strategy to navigate the geographical as well as social boundaries and a dis-cursive tool to fight their criminalization and marginalization by local authorities.

As revealed in this chapter, kinship and alliance relationships across the border were incorporated into the complex organization of border-crossings in a lucrative business, yielding profits from brokerage arrangements among large-scale entre-preneurs, tribal lineages as well as military and customs officers, who regulated these crossings. These relationships also informed manly cultures that forged a sense of attachment tested by loyalisms and betrayals of fellow townsmen joined in a smuggling venture. The discussion of this chapter shed light on the maneuver-ing capacities of poor dwellers in creating everyday tactics and strategies, which nevertheless reproduced the local power and inequality structures.

Notes

1 Öncüpınar border gate of Kilis has been closed to the crossing of Turkish nationals and vehicles registered with Turkish licenses in July 2012.
2 The unemployment rate is recorded as 10.1% in 2011. The number of Green Card holders is 31,656, the overwhelming majority being in the town center and the villages affiliated with the central district. The provincial director of the Central Employment Agency stated that their number recently decreased from 45,000. The number of Green Card holders climbed over the coverage rate of social security, being 19,458 dwellers in the town paying their insurance premium.
3 In a study of the transformation in the Turkish welfare regime, Ayşe Buğra and Çağlar Keyder states that the social assistance responses to the new poverty are prone to the clientelistic relations between citizens and political authorities, as "a number of largely unstructured and often traditionally rooted institutional arrangements define the area of social assistance" (Buğra and Keyder, 2006, p. 219).
4 "Tekel stopped its support purchases from farmers at advantageous prices," says Abdullah Aysu, chairman of the Farmers' Unions Confederation (Aysu, 2013). The statement of Kilis deputy Mehmet Nacar about the amount of fresh grape produce bought by the local Tekel factory in 1999 points to the public sector as the foremost buyer; see TBMM, 1999, p. 10.

5 See Kilis İl Çevre Durum Raporu, Kilis Governorship Provincial Directorate of Environment and City Planning (2011).
6 See Appendix for the names of state institutions interviewed.
7 According to my interviewees as well as testimonies of transport company owners, the Syrian government steered international trade between the two countries through Cilvegözü gate by indirect incentives such as avoiding charging extra fees and smoothing strict technical regulations in order to favor the Nusayri community at the border.
8 See the news report in Yeni Şafak (2000) and Aydın (2000, p. 309).
9 The Processing Regime came into force on January 1, 1996 via Decree No. 95/7615.
10 "Bürokrat Avı", *Hürriyet*, May 31, 2000.
11 "500 trilyonu aşabilir", *Hürriyet*, May 30, 2000.
12 In addition, it was reported that a banana export group complained about the unjustified profits (Yeni Şafak, 2000).
13 Serdar, Hamit, and Talip were young small-scalers who provided much information about cross-border trade.
14 *Bu iktidar geldi ilk Gümrük Yasasını çıkardı zaten. Bilginiz vardır. Gümrük Yasası çıktığı için şimdi bir kişinin yasal hakkı üç kilo çay, beş kilo şekerdir. Sigara falan yasaktır. Gümrük Yasası çıkmazdan önce zaten geçişler serbestti yani. Ben kendim boşaltıyordum arabayı. Bir ton şeker çıkartırdık arabadan.*
15 The interviewees calculated their gain as if the Turkish Lira was not redenominated by the time. The Lira was revaluated with the removal of six zeros in 2005.
16 *Tam seçim üstü birkaç gün çok sıkı oldu. Ben bunu gözümle gördüm; milletvekili geldi, Gümrük Meydanına. "Gümrük Müdürüne niye karakollar çalışmadı" dedi. Ben oradayım işte. "Bu kadar şeker getiriyorlar" falan dedi. Milletvekili elini kaldırdı, Gümrük Müdürüne tokadı indirdi. Ben bunu gözümle gördüm. Gitti dolandı, "görevini yapıyorsan yap, senin burada işin yok" dedi. O kadar halkın içinde adama tokadı vurdu, yerine gönderdi. Ondan sonra kapı açıldı serbest oldu. Muayene yok. Yani bir milletvekili istediği gibi kapıyı çalıştırabilir istedikten sonra.*
17 *Şimdi hocam, şeker bitti sigara başladı. O zaman kimse üç karton sigara getiremezdi. Şimdi elli karton getiren var.*
18 For example, Leyla, a woman married to and divorced from a Turkish man, still living in Kilis said that she was among the pioneers bringing cosmetic goods from Syria. When I interviewed her, she used to commute to Aleppo and peddled in the city streets in order to sell her Avon products that she carried from Kilis.
19 The Anti-Smuggling Law no. 5607 replaced the Law no. 4926, published in the Official Gazette no. 25173 on July 19, 2003. The new law appeared in the Official Gazette no. 26479 on March 31, 2007.
20 A carton of Winston cigarettes was sold in the arcade shops for TL 28 by summer 2011. The price of a pack in the regular market was TL 5.5.
21 *Adam getirmiş beş karton, on karton sigara; tutuyorlar, onu alıyorlar. Adam evini onunla geçindiriyor, yazık değil mi?*
22 *Kapı bir deve, bir fil. Yuttuğunu yutuyor, yutamadığını bırakıyor. Şirket hamuduyla götürüyor. Memur bizim gibi garibanları tutuyor, çalışıyorum oluyor.*
23 *Herhalde üç yıl önceydi, 2008'de üç günde birdi Suriye. Ondan sonra son iki yıl içersinde arabayla her gün giriş çıkış serbest oldu. Milletvekili seçimlerinden bir gün sonra haftada bire çıkarttılar arabayla. Arabayla haftada bir, yayan da her gün geçebiliyorsun.*
24 Decree no. 2011/2595, published in the Official Gazette no 28170, on January 11, 2012.
25 *Mesela biz günlük gidip geliyorduk. Yüz, yüz elli lira par kalıyordu günlük. Şu an gidiyorsun, biraz gazla atıyorsun üç yüz. Dört yüz lira para kalıyor. Her gün yüz elli kazansan haftada beş gün yedi yüz elli eder. Her gün imkan daha iyi idi. Getiriyordun satıyordun ediyordun, ertesi günü tekrar gidiyordun. Hiç yoktan uğraşımız oluyordu. Şimdi o da yok. Haftanın bir günü gidiyorsun, altı günü boşsun.*
26 By the summer 2011, a liter of oil was TL 4.23 in the regulated market.

27 *İki üç tane sigara yakalatsan bu önemli değil. Ama bir de ben arabamda gittim; beş milyarlık eşyayı tutup da sıfırlayıp seni mahkemeye verirlerse, onu bir senede çıkaramazlar. Mal yakalatmadan yakalatmaya fark var. O gün adam geldi benim üç beş kilo çayımı aldı git dedi. Bu bir şey değil. O gün benim arabamı muayene etti. Dönmedi, bir şey almadı. Papağan getiriyorlar oradan kirayla yüz liraya. Oradan alıyorsun buraya getiriyorsun yüz milyon para alıyorsun. Muayeneci oradan geçerken kuşun sesini duydu. Benim yüzüme baktı, ben bir şey demedim. Muayene etmişti. Geri döndü, geldi. Defteri yere koydu. Eğildi baktı, alttan kuşu gördü. Kuşu gördü ya, "bagajı aç" dedi. Bagajı açtım. Beş kilo çay aldı, götürdü ambara attı. Kuşu gördü ya, yoksa bir şey almayacaktı.*

28 *Krank var, karbüratör var. Adam diyor ki "şunu götür, al sana elli milyon, yüz milyon". Bunu da her adam alıp götürmüyor. Riskli. Onu Muayeneci görürse kesinlikle bırakmaz. Karlı iş ama riskli. Onu götürebilmek için ona göre arabanın yeri olmalı. Mesela, bizde kuşu gördü, elli liralık çayı aldı. Belli olmuyor yani. Biraz şans, biraz uyanıklık. Orada herkesin yapacağı iş değil.*

29 *Bakma sen! Akrabayız sonuçta.*

30 See van Bruinessen (2002) for his statement that the tribe members share an ideology of common descent, endogamy (parallel cousin marriage) and segmentary alliance and opposition.

31 *Akrabalar mal getiriyor ama komisyonla. Getirmediğimiz hiçbir şey kalmadı. Kilis'te bizi tüccarlar beklerdi; genellikle batıdan [gelirlerdi]. Akrabayla ortak da çalışılır. Paran yoksa akraba sermayeyi verir, sadece komisyonu alır. Ama sadece akrabalar yapar bunu.*

32 The signing of the Adana agreement in October 1998 was effectual in Syria's deportation of PKK leader Abdullah Öcalan and the water dispute eventually turned into a technical issue (Zafar, 2012). Scholars also argued that the changing state policy of the Kurdish issue towards desecuritization and democratization, particularly with the AKP's rule contributed to a shift in Turkish–Syrian relations toward cooperation and interdependence in securing regional peace (Aras and Polat, 2008).

33 This implementation was stopped in 2011 after the breaking out of Syrian unrest for reasons of security and I could not observe the crossing of the families in successive holidays though my prolonged stay in the town covered them.

34 During the year 2011, a Turkish Lira was about 32 to 36 Syrian pounds.

35 For example, Watenpaugh indicates that the newspaper *Halab*, a major publication of the early twentieth century promoting Arabic nationalism in the late 1920s, depicts "Turk" as "an ill-mannered individual, replete with onerous character flaws, intent on defying modern civilization, intellectually deficient and morally corrupt" (Watenpaugh, 2005, p. 10).

36 *[Türk gümrüğünde pasaport kontrolünden sonra] Onu alır, gideriz Araplara gireriz.*

References

Aras, B. and Polat, R. K. (2008) "From Conflict to Cooperation: Desecuritization of Turkey's Relations with Syria and Iran", *Security Dialogue*, 39(5): 475–495.

Aydın, M. (2000) "Operasyonun adı var", *Aksiyon*, no. 309.

Aysu, A. (2013) "Türkiye tarımının serbest piyasaya uyarlanması ve küçük çiftçiliğin tasfiyesi", in *Perspectives – Political analysis and commentary from Turkey*, issue no. 6 on "Grapes of Wrath – The Transformation of Agriculture and Rural Areas in Turkey", Heinrich Böll Foundation, Istanbul.

Baransu, M. (2000) "Paraşüt açılmadı", *Aksiyon*, no. 288.

Bedirhanoğlu, P. (2007) "The Neoliberal Discourse on Corruption as a Means of Consent Building: Reflections from Post-crisis Turkey", *Third World Quarterly*, 28(7): 1239–1254.

Bozçalı, F. (2014) "Hukuki-Maddi Bir Kategori Olarak Sınır: Türkiye-İran Sınırında Kaçakçılık, Mahkeme Süreçleri ve Sınırın 'Resmi' Temsilleri, *Toplum ve Bilim*, 131: 135–162.

Buğra, A. and Keyder, Ç. (2006) "The Turkish Welfare Regime in Transformation", *Journal of European Social Policy*, 16(3): 211–228.

Dinçer, O. B. et al. (2013) *Suriyeli Mülteciler Krizi ve Türkiye: Sonu Gelmeyen Misafirlik*, Ankara: USAK and Brookings Institute.

Doğruel, F. and Karakoç, J. (2013) "The Regional Repercussions of Turkey-Syria Relations", Athens: *ATINER'S Conference Paper Series*, No: POL2013- 0539.

Galemba R. B. (2012) "Taking Contraband Seriously: Practicing "Legitimate Work" at the Mexico-Guatemala Border", *Anthropology of Work Review*, 33(1): 3–14.

Galemba, R. B. (2013) "Illegality and Invisibility at Margins and Borders", *Political and Legal Anthropology Review*, 36(2): 274–285.

Gupta, A. (1995) "Blurred Boundaries: The Discourse of Corruption, the Culture of Politics, and the Imagined State", *American Ethnologist*, 22(2): 375–402.

Heyman, J. McC. and Cambell, H. (2007) "Corruption in the U.S. Borderlands with Mexico: The 'Purity' of Society and the 'Perversity' of Borders", pp. 191–217 in M. Nuijten and G. Anders (eds.), *Corruption and the Secret of Law: A Legal Anthropological Perspective*, Aldershot: Ashgate.

Hürriyet (2000a) "Bürokrat Avı", 31.6.2000. Retrieved 23 January 2018 from www.hurriyet.com.tr/burokrat-avi-39158267.

Hürriyet (2000b) "500 trilyonu aşabilir", 30.6.2000. Retrieved 23 January 2018 from www.hurriyet.com.tr/500-trilyonu-asabilir-39158012.

Keyder, Ç. and Yenal, Z. (2011) "Agrarian Change under Globalization: Markets and Insecurity in Turkish Agriculture", *Journal of Agrarian Change*, 11(1): 60–86.

Kilis İl Çevre Durum Raporu, Şehir Planlama ve Çevre İl Müdürlüğü, Kilis Valiliği, 2011.

NTV (2001) "Paraşütte yeni tutuklamalar", NTV, 8.5.2001. Retrieved 23 January 2018 from http://arsiv.ntv.com.tr/news/81588.asp.

Official Gazette no. 25173, 19.7.2003.

Official Gazette no. 26479, 31.3.2007.

Official Gazette no 28170, 11.1.2012.

Özden, Ş. (2013) *Syrian Refugees in Turkey*, Migration Policy Center Research Report 2013/05, San Domenico di Fiesole: European University Institute.

Parizot, C. (2008) "Crossing Borders, Retaining Boundaries: Kin-nections of Negev Bedouin in Gaza, West Bank, and Jordan", pp. 58–84 in S. Hanafi (ed.), *Crossing Borders, Shifting Boundaries: Palestinian Dilemmas*, Cairo: American University in Cairo Press.

Reeves, M. (2007) "Unstable Objects: Corpses, Checkpoints and "Chessboard Borders" in the Ferghana Valley", *Anthropology of East Europe Review*, 25(1): 72–84.

TBMM (1999) TBMM Tutanakları (Proceedings of TGNA), 21st period, 2nd legislation year, 7th assembly, p. 10, 14.10.1999.

van Bruinessen, M. (2002) "Kurds, States and Tribes", pp. 165–183 in F. A. Jabar and H. Dawod (eds.), *Tribes and Power: Nationalism And Ethnicity in the Middle East*, London: Saqi.

Watenpaugh, K. (2005) "Cleansing the Cosmopolitan City: Historicism, Journalism and the Arab Nation in the Post-Ottoman Eastern Mediterranean", *Social History*, 30(1): 1–24.

Yeni Şafak (1999) "Bayramlaşmada tel örgü engeli kalkacak", 15.6.1999. Retrieved from www.yenisafak.com/gundem/bayramlasmada-tel-orgu-engeli-kalkacak-586999.

Yeni Şafak (2000) "Bir İsim Bir Operasyon", 23.8.2000. Retrieved 23 January 2018 from www.yenisafak.com/arsiv/2000/agustos/23/g3.html.

Yeni Şafak (2000) "Gaziantep'e İnen 'Paraşüt'-1: Muzun dayanılmaz cazibesi", 6.1.2000. Retrieved 23 January 2018 from www.yenisafak.com/arsiv/2000/haziran/06/dizi.html.

Zafar, Shaista Shaheen (2012) "Turkey's 'Zero Problems with Neighbours' Foreign Policy; Relations with Syria", *Journal of European Studies*, 28(1): 143–158.

Conclusion

When I decided to go to Kilis town in order to conduct fieldwork at the border setting, I wanted to have a story to tell to as a graduate student of sociology in return, let alone a substantial dissertational study. Indeed, I had a story after I finished the field research. I could talk for hours to anyone who was asking about Kilis how I strove to enter 'the field' and managed to penetrate the secretive world of locals, making them talk about how they survived mine explosions, how they suffered from the militarized border security compelling them to issue a permission even to labour on their own land at the border, how they could smuggle so many goods at once in the hiding-places of a car, how they had a quarrel with the Syrian women that their husbands brought from Aleppo as second wives. My field story is not only a physical journey to the border, but also a cognitive one because of their stories. In this border town, everyone had their stories to tell. Elder members of notable families, village men, widow women, even high school students, anyone can give a vivid testimony about how the border impinged on their daily life without even being aware of it. I believe that, in all these stories, one can find how town dwellers make themselves the protagonists of their own stories as they come to grips with various contexts set by the border.

In this book, I try to interpret how their stories turn 'extraordinary' situations into normal ones and, thus, give border its historicity. The discussion of this book has problematized the notion of border as enabling mobilities and producing closures, while pointing out that the border inflicts subjectivities, values, and practices in a way that is not found elsewhere in the nation-state. I have indicated the ways in which socio-economic strata manipulated and circumvented the territorial, cultural, and legal boundaries and shifted the meanings of illegality, wealth, and work. Here I want to develop some evaluations.

My field research has revealed the ways in which the border dramatically shaped subjectivities of Kilis dwellers, extending its impact to all strata including the elites. For instance, Kilis dwellers were stigmatized in various ways: notable families as usurpers and traitors, new wealth as crime bosses, and the urban and rural poor as lazy and having the habit of making money with little effort. Moreover, the infamous identification of Kilis town with illegality and disorder affected every local, particularly those settled in the metropolitan as well as Anatolian cities. In Chapter 4 on new wealth, I have pointed to the delicate balance between secrecy and transparency that the Kilis migrants in Istanbul try to keep in order to ensure community support and sustain their image of decent businessmen. Philanthropic activities are a major source of legitimization for these businessmen. Not all businessmen have a high profile, unlike the barber, a former smuggler who was implicated in gold smuggling, money laundering, and fraudulent export in the 1980s and eventually gained social recognition as a decent businessman after he had been released of all charges. The new wealth, who since the 1960s have turned their entrepreneurial ventures in Istanbul into stable businesses, are well-known in their hometown for their charitable activities. But they tend to keep a low-profile otherwise.

As reflected in the barber's words, comparing his self-made success with the two richest capitalists of the country, Koç and Sabancı, the new wealth of Kilis is defensive and alert. As a reminder, the barber stated that he cannot be labeled as being a heroin and arms dealer, since he had grown rich out of barbering in the same way as Koç and Sabancı rose to wealth from modest economic backgrounds. As stated in Chapters 4 and 5 on new wealth, the field study revealed that the gold trade during the 1960–1980 period created an enrichment among families from rural backgrounds, as the nicknaming of enriched locals as "rich" illustrated. However, as the case of the barber illustrated, the new wealth tended to obscure the roots of their wealth. For instance, the barber put forward his humble roots as a barber rather than highlighting his financial achievements.

As a major finding of the study, the transformation of the Ottoman inland frontier into a modern nation-state border has had a dramatic impact on the social stratification structure in Kilis town. The social mobility processes in Kilis town critically diverge in terms of subjective perceptions and evaluations as well as local structures of opportunities, though they could be analyzed in conformity with regular patterns such as intergenerational occupational transmission, the role of marriage, access to education, and housing as symbolic capital. The in-depth interviews with Kilis locals offered four broad stories of social mobility that I have organized into three parts. These stories reveal in what ways the local structures of opportunities bring forth class sentiments (and injuries) among various socio-economic strata and the cultural and economic capitals of the family are transmitted between generations. They also indicate that Kilis locals adopted navigation across the territorial, cultural, and economic boundaries as social mobility strategies, which both secure the social reproduction of status and lineage and result in the abrupt changes in social stratification structure.

The first story concerns the decline of notables. The notables, including both traditional landed families and trade families, composed the old wealth of the town and they experienced a feeling of downward mobility even though they dominated economically, politically, and culturally until the 1960s. The label of notable, that is, *eşraf,* is ascribed by the locals as well as notables themselves only to the traditional landed families for historical and normative reasons. The book showed that the sense of falling from grace is particularly experienced among *eşraf* rather than *esnaf;* that is, trade notables. The interviews indicated that the traditional landed and trade notables competed with each other in political and cultural terms.

Politically, as traditional landed notables demanded that Kilis should stay annexed with Syria and abstained to a large extent from getting mobilized in armed gangs fighting against the French occupation, they were stigmatized as traitors and faced the danger of losing their entitlements to basic citizenship rights. Thus the traditional landed notables were forced to share their dominance with the trade notables which gained power during the Independence War against Allied countries. Culturally, the traditional landed notables competed with trade notables in the field of culture and consumption. Although the trade notables could more easily adapt Westernized tastes and lifestyles, as they lacked the deep attachment to lineage and patrimony prevalent among traditional landed families, the latter did not lose their status. The paternalistic relationships based on large landholding and patronage largely thwarted the conversion of economic capital held by trade notables to cultural capital and achieved the social status of traditional landed notables. However, the gradual introduction of the money economy as of the 1950s and the rise of new wealth in subsequent decades changed the situation and culminated in the feeling of decline among traditional landed notables.

The interviews also indicated how the traditional landed and trade notables coped with their feeling of decline: seeking access to education and urban employment, migrating to cities, and seeking alliances with the new wealth through marriage are the principal means of social reproduction among traditional landed notables. Although the traditional landed families also embraced illegal trade as much as trade families did, the sense of decline was too strong among the older generations of traditional landed notables because, this study showed, they had to embrace what was once disgraceful for them: that is, circumventing the law in order to engage in illegal trade.

The second and third stories concerning the rise of new wealth as well as the rise of extended families from a rural background among the urban middle class in the 1960s can be read together. The study demonstrated the growth of the shadow economy in Kilis, which enabled upward mobility to the new wealth and several extended families from rural backgrounds as they relied on paternal relations of patronage and kinship. I adopted the term shadow economy in order to describe various sorts of income-generating practices that are not regulated along state and market channels. These practices encompass legal, semi-legal, illegal, and informal activities and they are enabled by transnational networks operating outside the state institutions, although in complicity with the local authorities and state bureaucrats.

The interviews indicated that the local and regional markets of gold, foreign exchange, and consumer goods at Kilis border were connected with financial and trade centers like Beirut, Istanbul, and Zurich as well as with transnational crime organizations long before the neoliberal trade liberalization after the 1980 coup. These economic practices partially constituted a redistributive dynamics providing income and employment for town dwellers. But the emergence of family monopolies around their enterprises resulted in equal distribution of the profits reaped from illegal trade and led to an economic accumulation. This economic accumulation was initially criminalized by hegemonic discourses of politics and media but it was socially accepted later on as rightfully earned wealth as a consequence of neoliberal policies to finance the export-led growth on the basis of the 'economic punishment for economic crime' principle. It was in this context that new wealth and extended rural families experienced their rise.

The shadow economy also undermined the traditional industry based on agricultural production and artisanship, changed the occupational composition in the town, and altered the urban landscape. However, it also strengthened relationships of patronage and kinship and contributed to the emergence of new authority figures in Kilis. Local chieftains and heads of extended families that became rich as a result of illegal trade assumed their authority by imitating the old notables as well as contesting them. They employed dwellers from the poor strata by engaging them in patronage relationships in which the entrepreneurs patronized and dominated the poor by demanding their labour not only in carrying their goods across the minefields but also looking after their errands. They also exercised political influence over them and recruited them as potential voters for the political parties they supported.

The fourth concerns the transformation of the rural and urban poor into border labourers. The petty-commodity production in agriculture was not enough for the rural poor to subsist in Kilis and they were largely dependent on landholders holding large-scale lands. The large-scale landholding and sharecropping arrangement between landlords and peasants started to decline in the 1950s and illegal trade across the border yielded small gains for the peasants. However, while the growth of the shadow economy liberated them from the paternal relationships of patronage established with large landholders, they quickly became dependent on large-scale entrepreneurs, which relied on tribal chieftainship and kinship. The new patronage relations provided the lower strata with protection from law enforcement and alleviated the risks of criminalization as well as physical injury across the minefield.

The shadow economy declined after the 1980 coup and large-scale entrepreneurs, namely the old and new wealth, moved to cities like Istanbul, Ankara, and Gaziantep. But it was revived by the cross-border trade introduced by the government in the mid-1990s, which allowed semi-legal small-scale trade. The lower strata commuted across the border almost daily in order to earn their living—and they did this until the Turkish government closed the border gate with Syria in July 2012 for security reasons. This small-scale trade yielded an extra income for them, sometimes equal to their salary. But in the absence of patronage by large-scale entrepreneurs, they were increasingly exposed to the dangers of criminalization and punishment by law enforcement and being unable to pay criminal fines, dangers of growing economic indebtedness. Remittent profits during the neoliberal period made the urban and rural poor completely dependent on small-scale trade. The interviews showed that border commuters from lower strata dwellers were aware of the unequal distribution of profits from cross-border trade, even though it was a specific trade regime regulating the trade activities presumably under equal terms for all the traders. Lower strata commuters could not afford to bribe customs and cover the damages of getting caught or losing their goods. Since they tended to individualize the risks in the absence of patronage, the profits yielded from small-scale trade were unequally distributed among the poor as well. Still, small-scale trade during the 1970s and in the last decades helped the urban and rural poor to establish themselves as protagonists of their own 'success stories.' My interviewees, both male and female, boasted about establishing themselves as patrons who made their own money.

These stories point to the ways in which the border context shifted the meanings of illegality, wealth, and work as well as shaped various cultural and legal belongings. These meanings also varied among different socio-economic strata. For instance, embracing the illegal means of economic accumulation was once construed as disgraceful for the traditional landed notables because they relied on the lineage and patrimony of their ancestors. Wealth for them was not achieved, but it was ascribed by status. Thus, although they were engaged in illegal trade in order to reproduce their social standing, they felt it as losing their proper place. On the other hand, the old wealth were not stigmatized as lawbreakers as the new wealth were. The new wealth lacked the social capital to establish recognition and respect as notables and they were criminalized. So they needed to restore their image as philanthropist businessmen, imitating the old wealth by taking part in charitable activities in the town and turning their economic accumulation into socially accepted wealth.

For the middle and lower strata, engagement in illegal trade replaced regular employment and it was normalized as work. For instance, the smugglers' sharia which determined the codes of conduct for illegal trade imitated a guild organization and therefore established smuggling as an occupation. The interviews also indicated to the intergenerational transmission of smuggling as an occupation as a normalized pattern of social mobility. However, these strata also gave consent to the status quo, by assimilating that the means to their living were unlawful, although they saw it as legitimate and they tended to individualize the risks of criminalization and stigmatization by state policies.

These accounts revealed how Kilis locals dramatize themselves as protagonists of their own 'success stories.' I tried to discuss in detail how the local artisans and sharecropper peasants were emancipated from paternal relations based on agrarian production and recruited by large-scale entrepreneurs as border labourers. Nevertheless, the town dwellers boastfully asserted themselves as "patrons" although profits reaped from the shadow economy were unequally distributed between upper and lower strata as well as among the poor. The rural and urban poor could yield extra earnings and, in some cases, they managed to compensate for their low salary with trade gains. Also, the fact that remittent profits from the shadow economy were possible kept the expectations for earning a decent living from the border alive. Nevertheless, as the discussion of Chapters 6 and 7 on the rural and urban poor showed, the prospects of upward mobility for the poor were likely to be hindered by structural constraints and short-term gains might easily be swept away by arbitrary policies of border and customs control as well as international tensions.

The border context underlined new mobility opportunities as the town dwellers creatively appropriated in different historical periods the governmental policies that sought to regulate or ban trade relations. The border dispute and negotiations with the French Mandate in the 1920s resulted in a specific border transit regime called Passavant, allowing the entitled landholders to labour on their land and harvest their crop. So the Passavant regime promoted the vested interests of traditional landed notables and allowed many notable families in Kilis to sustain land tenure in Syria. But, the Passavant regime also allowed notables to capitalize on the regime to yield differential revenues from the selling of their produce in foreign markets, although they were supposed to bring their produce along the Turkish borders for their own consumption or selling in the domestic market. The border-crossing regime also allowed them to engage in gold trade.

On the other hand, strong families in rural areas took advantage of high tariffs imposed by the French-Mandate in order to sell several contraband items on the black market. The breaking up of southern Anatolia from Aleppo did not prevent the perpetuation of inland trade between Kilis and Aleppo, but turned it to an illegal trade criminalized by the state authorities as early as the 1930s. Rural families smuggled big flocks of sheep and cattle and olive oil to northern Syria, while they brought contraband items such as coffee, salt, flint, European fabric and tailored jackets, gas oil, and sugar.

Town dwellers manipulated the protective measures of the import-substituting industrialization period of 1960–1980 by reckoning rents to the illegal entry of consumer goods as well as gold and foreign exchange. Neoliberal policies further allowed rent-seeking enterprises to maneuver and reap high profits. Since the 1980s, the traders benefited from various state subsidies to promote an export-led growth in the countries. Although these practices were associated with the Özal period when the Prime Minister Turgut Özal headed the government in the 1980s, they continued in different forms through the 1990s, this time making the state itself an actor manipulating its own regulations and blurring the boundaries between legal and illegal realms. Town dwellers continued to circumvent for instance the regulations of cross-border trade, a specific customs regime allowing border cities to benefit from tax reduction or exemption in import or export.

Although Turkish and Arab nationalisms are strong at the border, the study showed that Kilis dwellers navigated these boundaries by relying on cross-border kinship and alliances between families. However, the meanings of these relations were ambivalent and changing across different socio-economic strata as well. The interviews demonstrated that the notables had great difficulties in claiming their Syrian connections and kinship relations because of their stigmatization as traitors. Most of their relatives were left in Syria because of the political campaign run against the absentee notables who left Kilis for Aleppo and who did not get mobilized in national gangs during the French occupation. Thus when notables claimed their kin relations, they tended to emphasize their Turkish identity or their contribution to the Independence War, testifying to their nationalistic sentiments.

The cultural boundaries among notables were likely to be undermined only with reference to the past, revealing a nostalgic attachment to the home town, which was imagined as a geographical extension of the city of Aleppo. As the Passavant regime promoted cross-border movement and the maintenance of strong patrimonial ties with Aleppo and its countryside, the border did not constitute a territorial barrier for them until its mining and the confiscation of their landed estates by the Syrian government. In-depth interviews with notables tended to involve a nostalgic return to the past and evoked their memories of farm and social life in rural Aleppo or the city center. Notables depicted Kilis town as a civilized, modern, and cultured place because they associated it with the city of Aleppo, a modern and cosmopolitan city integrated with world capitalism at the late Ottoman era.

Cross-border kinship and marriage alliances proved to be resilient and versatile ways of sustaining and normalizing illegal exchange practices for the middle and lower strata. Extended families used it as an asset to maintain illegal trade. Since without a legal framework, the risks of being cheated were high and traders had to rely on their bonds of trust and affinity. Extended families used kinship in order to keep their enterprises as family business, particularly through cousin marriages, which indicated the persistence of tribal lineages. Kinship relations sustained by loyalty to a chieftain, though weakened in Kilis, helped extended family members to rise as an authority figure with the power to make profit-sharing agreements with military officers.

However, the book also demonstrated that the open-border policy of the Turkish government, especially during AKP rule, reinforced the perception of difference and social distance, although it facilitated commuting across the border and created a symbiotic economy among the border communities at both sides. Local reactions to the transfer of Syrian migrants from Hatay to Kilis in early 2012 was illustrative. Local business and shop owners in particular strongly opposed their transfer. Even the news anticipating their coming were enough to evoke negative stereotypes and images about Arab identity and reinforced cultural boundaries between communities. That is, locals of Kilis tended to disregard the kinship relations when they believed that their means of living was threatened.

But the poor employed kinship relationships and cross-border marriage as a material strategy to navigate the geographical as well as cultural boundaries and a discursive tool to fight their criminalization and marginalization by local authorities. The poor commuters relied on their kin as business networks and for temporary stays across the border. So, these relations helped them to normalize their border crossings in the eyes of local authorities.

I underlined in this book the resilient and versatile strategies adopted by Kilis dwellers in coping with various types of stigmatization and criminalization. Town dwellers meticulously crafted relations of reciprocity, alliance, and kinship in order to sustain their living under the conditions that threatened their everyday routine. The discussion of this book accentuated how dynamics and processes at the border may easily shatter their lifelong efforts to build a stable life. As discussed in the preceding chapters, illegal or informal practices may partly replace redistributive mechanisms in complicity with state authorities in neoliberal regimes, as illustrated by the cases of the erupting informal sector with transition to a capitalist economy in post-Soviet countries (Pelkmans, 2006), emergence of local agrarian economies and illegal trade in Latin America within the context of multinational free trade agreements like the North American Free Trade Agreement (NAFTA) (Galemba, 2008, 2012a, 2012b), illegal trade and gang-based road banditry in African borders (Roitman, 2004, 2006), and professionalization of the informal economy of border crossings at Israeli checkpoints (Parizot, 2014). In my view, the lack of mistrust or unrest among the AKP voters suggests the normalization of illegal or informal economic practices as redistributive mechanisms with the transition to the neoliberal regime turning the right to a secure salary into a privilege, undermining peasantry and promoting a welfare regime based on assistance dependency and political patronage.

The case of Kilis demonstrated how dwellers sought to determine their own destiny rather than expecting solutions from the government at a locality where the state failed to promote public and local entrepreneurship and provide employment to the town community. It has also underlined that illegal practices cannot necessarily be considered as wrongful because they are regulated by an ethics drawing on local power structures, state regulations, the code of smuggling as well as religious values. I suggest that the case of Kilis can inspire one to think about the post-1980 cultural transformation boosting the pragmatic ideals of 'turning the corner' associated with the Özal legacy and give insights into the question of economic justice, revealing the lower strata's income-generating practices.

Furthermore, the case of Kilis highlighted the ways in which local power and inequality structures are reproduced by the everyday, oblique, and improvised strategies of dwellers and demonstrated that the 'traditional' structures facilitated transnational networks of capitalism. My discussion demonstrated the continuing social and economic significance of local notables in the early Republican regime as an aspect of social stratification structure inherited from the late Ottoman social order. As detailed in Chapter 2 on local notables, these families sustained their power until the 1960s, even though they had to share it with a new middle class, which emerged within the context of rising Turkish nationalism during World War I and local mobilization against the occupation of Allied forces. It is demonstrated that notable families not only maintained their status on the basis of their lineage and patrimonial heritage, but also dominated public life upon the paternalistic organization of agrarian production. The persistence of the social and economic power of local notables has some implications for scholarly literature on Turkish modernization and capitalist development. Yet, while recent studies on local notables address the persistence of local notables in cultural and political terms, as in the examples of Karadağ (2005) and Meeker (2002) cited by this book, they do not address the economic repercussions.

The Keyder vs. Köymen debate on the capitalist relationships originating in large landholding largely excluded the role of border at the southeastern margins of the newly established Republic. As reviewed with regard to Keyder's main arguments, he carefully delineates the exceptional characteristics of the Ottoman Middle East regarding the conditions of agricultural production. He suggests that even though the landlord–peasant relationship was an exploitative one, landlord-managed estates did not tend to export their agricultural produce. The findings of the study seem to contradict this by pointing to the ways in which notable families yielded differential profits from agricultural and gold trade by manipulating and circumventing the border transit regime and customs regulations. While Keyder admits that well-placed bureaucrats seized new opportunities of the circulation of products and money, I extend this argument to the whole group of local notables in Kilis.

Furthermore, I suggest that the perpetuation of power of these families was remarkably effective in the transformation of the regional economy into a lucrative cross-border trade from the 1960s onwards, as they pioneered capitalizing on the border transit regime. The perpetuation of their power is also significant in terms of relations of patronage. As argued in the chapters on the rural and urban poor, the perpetuation of the relation of patronage until the 1960s, disguised in the form of charity and gift economy, prepared the conditions for the rise of chieftains or heads of extended families, muhktars as well as peasants who became rich with illegal trade as authority figures. These authority figures both contested and imitated local notables in maintaining paternal relations with the rural poor and patronizing them as the latter earned their livelihood as hired porters carrying goods across the mined border. Nevertheless, these patronage relationships between the authority figures and peasants differed from landlord–peasant relationships in that the former was governed by the money economy.

I aimed to contribute to border studies within the Middle Eastern context by illustrating culturally informed strategies of border dwellers in accommodating various boundaries set by the nation-state. The case of Kilis underlined the perpetuation as well as reinterpretation of paternalist and patriarchal relations within modern contexts. The narrative accounts of town dwellers also construe the border life in gender terms. I believe that visual media and literary texts such as movies, documentaries, and literature also convey meanings and images about border, which inform popular imaginations about life at the border. It is not surprising that the social realism tradition in Turkish literature and cinema[1] told the plight of poor male peasants at the southeastern borders of Anatolia, who are injured, maimed, or exploited by the village chieftains while carrying smuggled goods across minefields. For instance, the well-known movie *Hudutların Kanunu* (Law of the Border), written and directed by Ömer Lütfi Akad in 1966, portrayed Yılmaz Güney as the protagonist of the story, a poor peasant who is caught between an oppressive local agha and a military officer committed to terminate smuggling and who ends up dying in the minefields. There is only one female character in the movie, a young school teacher from an urban middle-class background, who tries to dissuade Yılmaz Güney's character from smuggling.

Similarly, narrative accounts in Kilis depict border life with reference to the plights that men have suffered in order to survive. Women can tell their stories about early marriage, domestic violence, leaving their parental home, being unable to get along with their Syrian daughter-in-law, managing a living after the spouse's death, taking care of the grandchildren after the fleeing of their mother with another man but these stories do not dominate the narrative construction of their identities as Kilis locals. The manly world of smuggling further marginalizes their stories as women.

To conclude, I argue that the discussion of this book may inform novel methodological approaches elaborated here about the integration of Middle Eastern societies with global processes, by focusing on how border communities come to grips with new border contexts. Situating their perspective on Middle Eastern borders, researchers can learn from intriguing strategies devised by ordinary citizens to maintain their living as well as 'hidden mechanisms' informing these strategies that the state may not register.

Note

1 Social realism, focusing on the problems of social classes and aiming at a realistic representation of society, has become influential in Turkish literature and cinema since the 1960s. The short stories of Şevket Bulut, including *Kaçakçı Şahan* (2011) [1970], the story of a peasant smuggler, which gives the book its title, exemplifies how border life and the plight of poor border dwellers became a noteworthy topic for social realist authors.

References

Bulut, Ş. (1991) [1970] *Kaçakçı Şahan*, İstanbul: Everest Yayınları.

Galemba, R. B. (2008) "Informal and Illicit Entrepreneurs: Fighting for a Place in the Neoliberal Economic Order", *Anthropology of Work Review* 29(2): 19–25.

Galemba, R. B. (2012a) "Taking Contraband Seriously: Practicing 'Legitimate Work' at the Mexico-Guatemala Border", *Anthropology of Work Review* 33(1): 3–14.

Galemba, R. B. (2012b) "Corn Is Food, not Contraband": The Right to "Free Trade" at the Mexico-Guatemala border", *American Ethnologist*, 39(4): 716–734.

Karadağ, M. (2005) *Class, Gender and Reproduction: Exploration of Change in a Turkish City*, unpublished PhD thesis, University of Essex.

Meeker, M. (2002) *A Nation of Empire: The Ottoman Legacy of Turkish Modernity*, Berkeley: University of California Press.

Parizot, C. (2014) "An Undocumented Economy of Control: Workers, Smugglers and State Authorities in Southern Israel/Palestine", pp. 93–112 in L. Anteby-Yemini, V. Baby-Collin and S. Mazzella (eds.), *Borders, Mobilities and Migrations Perspectives from the Mediterranean 19-21st Century*, Brussels: Peter Lang.

Pelkmans, M. (2006) *Defending the Border: Identity, Religion, and Modernity in the Republic of Georgia (Culture and Society after Socialism)*, University of Cornell Press.

Roitman, J. (2004) *Fiscal Disobedience: An Anthropology of Economic Regulation in Central Africa*, Princeton University Press.

Roitman, J. (2006) "The Ethics of Illegality in the Chad Basin", pp. 247–272 in J. Comaroff and J. Comaroff. (eds.), *Law and Disorder in the Postcolony*, University of Chicago Press.

Epilogue
War spillovers on Kilis borderland

Despite the strong convictions among townspeople and their prospects about cross-border trade, the armed conflicts escalating throughout Syria halted the cross-border movement unilaterally. The Syrian conflict kicked a refugee movement in the reverse direction, Kilis being turned into a new hub for transit. The abrupt halt to small-scale trade and other forms of cross-border exchange undermined the livelihoods and compelled the town inhabitants to cope with the new border context anew. The previous chapters of this book portrayed in detail this persistent endeavor of *dwelling on the border*. The Syrian conflict constitutes a landmark in the life cycles of this borderland, placing it in the midst of global changes in international relations, warfare, and humanitarianism. The policies introduced by the Turkish authorities to tackle mass displacement and criss-cross movements at the border produce new rebordering and debordering processes magnifying the changes in border settings.

Sarah Green likens such dramatic changes to an earthquake that rearranges the landscape (Green, 2012). These brisk changes align long-established ties with new actors, rework former border arrangements with new capacities and channel new flows through old routes. Despite the fact that the Turkish government closed the border gates, Kilis border remains as porous as before. This porosity, however, does not suggest the mere continuation of the *status quo* at the border. The new border porosity is twofold: the retreat of the Syrian state from the border is complemented by the strong intervention of the Turkish government aiming at a buffer zone and geopolitical control at the border area. Hence, this epilogue, aiming at a snapshot of war spillovers on the Turkish–Syrian borderlands, sheds light on the emergent forms of movement and enclosure shaping the sense of border among the overpopulated inhabitants of Kilis town.

Shifting demographics in the town

The intensification of conflict at subsequent stages of war has resulted in the fleeing of civilians from Syria en masse and seeking refuge across the Turkish border. The total number of registered Syrian refugees in Kilis reached 129,221 in 2016, of whom about 35,000 are camp settlers (Biehl et al., 2016). The internationally acclaimed "open-border policy" of the Turkish government sought at the beginning of the war to offer temporary refuge to fleeing Syrians with the anticipation that Assad's fall would be quick and refugees would return to their home. Hence, the reception of refugees is guided by a discourse of hospitality that welcomed the Syrian guests in camps set up in the border zones. Two camps, with containers offering shelter space to the refugees of 21 square meters, were successively opened in Kilis on March 2012 and June 2013, following the refugee amassments at the border entries. Öncüpınar camp was converted from an old camping ground on the buffer zone between Turkish customs and the Syrian border. Öncüpınar has been put to use as a center with a capacity of 2,053 containers, while 3,592 containers have been settled in Elbeyli camp a few hundred meters to the borderline at the eastern lowlands of Kilis (Ay, 2014). Facing ever-increasing demand, the sheltering capacities of the two camps in Kilis would have been increased or excessed in the upcoming years.

Camp refugees are not the main factor behind the skyrocketing numbers in Kilis. Self-settlement among refugees in the urban centers is a historical and ascendant tendency (Chatty, 2017). Presently, Kilis is known as the Turkish town with the highest ratio of Syrian refugees to the local population. The locals allege that the refugees in the center outnumber them.[1] Although the records of registered refugees do not specifically classify their places of residence and merely distinguish between camp and non-camp refugees in the town, refugee concentration in the urban center, together with the everyday in-and-out movement of refugees in Öncüpınar camp, at a close distance to the center, gives rise to the heightened presence of refugees and local anxieties about them.

AFAD prides itself on the "high rank life standards" offered in the camps, with health, education, banking, and telecommunication services plus other social facilities available to the refugees beyond the well-provided minimum requirements such as food, shelter, and camp infrastructure (Erkoç, 2013). The Turkish border camps absorbed the early refugee arrivals and appeased the humanitarian crisis to a degree. Whether the camps consist of a response to the crisis is another matter. I assume that the much-lauded living conditions of Kilis camps rather showcase the Turkish policies of refugee reception, which are positively appraised by the international humanitarian community (McClelland, 2014). The mass displacement created by the war in Syria rather induces the governments in neighboring countries, lacking an asylum framework, to pile the burden of welcoming refugees and dealing with the uncontrolled population growth on the shoulder of locals, particularly in the destitute border regions (Carpi and Şenoğuz, 2018).

A significant ratio of refugees moving to Kilis found temporary shelter in their kin's home (Özden, 2013). The house-owning locals opened up their property to their incoming kin for a small rent or as an outright gift. As the following section details, the gift-giving character of the hospitality relations was also revealed in the welcoming of refugees in early arrivals by the locals who attempted to provide them with aid, giving staple foods and basic furnishings and donating to the humanitarian campaigns. However, the local efforts were quickly overwhelmed by the constant refugee movement, especially in the town center as shelter as well as transit. It is estimated that the incoming of Syrian refugees to the town spurred a population increase of 98% (Biehl et al., 2016). Because of the economic dire straits in the countryside, the urban conglomeration in the town center has long been a pull factor for internal migrants. Thus, the overwhelming majority of the town population used to live in the small urban area and the population density already gives the local state hardship in providing urban services. The town center absorbs refugees, as many as locals, a fact which puts higher pressure on the public sector. Although Kilis soon became a focal point for international and national humanitarian organizations, the public sector is compelled to run their services without personnel and financial back-up.

The Turkish government has adopted a temporary protection framework for Syrian refugees since 2014 and supports their differential inclusion (Baban et al., 2016) to the society—granting them social citizenship rights and yet excluding them from asylum seeking. Turkey is party to the 1951 Geneva Convention on the Status of Refugees but maintains a geographical limitation, making it impossible to lodge asylum claims for refugees coming from outside Europe. The registered refugees in Kilis can benefit from public health services when they provide their identification documents—temporary IDs. The packing of refugees into the corridors of public hospitals, which are already at full working capacity for the treatment of wounded warriors of opposition groups transported across the border, raised complaints among the town locals who felt eclipsed by them in accessing health services.[2] The challenges in regulating increased motor traffic, also heightened by the lack of control over Syrian-licensed vehicles, were raised. The scarcity of housing in the town not only caused the overpopulation of individual tenements but the bus terminals, empty lots, abandoned buildings, parks, and mosque yards also hosted makeshift refugee tents (Deniz et al., 2016; Yaşar, 2014). These grievances often turned into tensions between locals and refugees, which prompted the local non-governmental organizations (NGOs), in a humanitarian disposition, to voice their concerns about increased burdens on the town in the face of the population upsurge in their 2013 report.[3] They demanded the declaration of a disaster situation which requires the allocation of an emergency budget to alleviate the emergent problems in public services including judiciary, health, social aid, road maintenance, and preservation of parks and green spaces.

The encounters between the locals and refugees is also impregnated by pre-existing ethno-religious tensions. The last section of chapter 6 illustrated the revival of negative stereotypes and images of Arab identity in Kilis with the news of incoming syrian refugees. The perception of Arabs is entangled in this region with the indoctrinated war accounts popularizing the local militia resistance to the Anglo-French occupation during the years 1919–1921 at the end of World War I. Despite its cosmopolitan background and ties extending across the border, Kilis is a stronghold of Turkish nation-building, clearly demarcating lines between Turkish and Arab culture. The Syrian displacement has significantly challenged this status quo as well as altering the minority status of local Arabs. While the refugees are not classified by their ethnic background by state registers, it is estimated that the Arab population in Kilis swelled from 1% to 37% of the population with the incoming of refugees (Cagaptay, 2014). Hence, the refugee perception in Kilis amplified the cultural differences with the Syrians.[4] It is reported, for instance, that talk among the locals stigmatized the displaced Syrians, especially living in makeshift settlements as "backward," "dirty," and "bad" people (Yaşar, 2014, p. 51).

A gender dimension to these tensions is revealed by the local anxieties about cross-cultural marriages between local men and Syrian women. The town has become the locus of increased rates of divorce among the local population, enticing experts to speculate about its causes—marriage to Syrian women (Orhan and Gündoğar, 2015). Reports estimate a growing trading on marriage based on sexual exploitation of Syrian women married to local men through informal arrangements (Dinçer et al., 2013). Since polygynous marriage is not legally allowed but socially licit in certain regions of Turkey, the marriage arrangement is performed according to religious customs but by a mosque *imam*, an appointed official. This sort of arrangement, which could easily be renounced by the husband's wish to divorce, serves as a cover for trading in marriage and forced prostitution. It is reported that these marriage arrangements, including early age marriages, were undertaken by organized crime networks located in the town and even extended to the refugee camps and camp officials (Mazlum-Der, 2014; ECPAT, 2015). While feminist scholarship stresses the gender and sexuality imbalance imposed by these marriages (Kıvılcım, 2016), there are also statements on the basis of refugee accounts pointing to the survival strategies sought by refugee women through these transactions (Herwig, 2017). Yet, it is worth admitting that the complex cross-border alliances and war economy blurred the distinction between the transactions as survival strategies and as organized crime.

Welcoming the Syrian refugees

Although border cities harbour strong grievances about the refugees, the authorities publicize local hospitality as a way of tackling the humanitarian crisis. Kilis stands out among other locales as the first border town to be nominated by the AKP deputy chair responsible for human rights as a candidate for the Nobel Peace Prize for its exemplary work in welcoming high numbers of refugees, despite the constraints on the livelihoods and material infrastructure caused by the population growth (Hürriyet Daily News, 2016). Kilis mayor Hasan Kara, former AKP parliamentarian run for two times, explained the motive of the candidacy campaign as the desire to offer the town as an altruistic model forging a peaceful cohabitation between two communities. Kara as the campaign spokesperson pointed out the limitations of local hospitality as well. On the eve of a controversial European Union (EU)–Turkish deal,[5] he made clear his intentions to urge European partners, especially Germany as the leading enforcer of the deal, to procure aid for the little town in order to rehabilitate the infrastructure as well as help the Turkish nation in providing decent living conditions for refugees.

The mayor Kara initiated an actual invitation to the German President Angela Merkel, which put the little town under the spotlight of Deutsche Welle, the German public international broadcaster. Kara's invitation aimed advocacy for Kilis'

Nobel nomination at her. Making it a spectacular event, Kara had led the preparation of an invitation to the German President signed by Syrian and Turkish women (CNN Türk, 2016). The ceremony in the main square of the town filled by women carrying placards of Frau Merkel was attended by Guiness' Turkey representative because the 1.5-m-wide and 19-m-long invitation constituted a World Record attempt.

As I fixed for Deutsche Welle staff a visit to Kilis town to follow up on the refugee situation, I had the chance to observe how the Mayor, in his rather populist performance, strived to craft an image of his people as generous and, most of all, his country as welcoming. Completing a visit to the municipal centers where the Syrian women and children enjoyed training and social activities, he escorted his visitors through a walk in the town streets, which took in several stops to the inhabitants and shopkeepers—both Turkish and Syrian nationals—on the way, as he shook hands and exchanged greetings frequently. In a sense, his comfort in walking across the town, and ease at communicating not only attested to the social cohesion and safety prevailing in the town, but also confirmed his self-assurance in his statements about the townspeople's altruism. He considered the latter as a lesson of humanity to the world.

A growing literature on the contemporary displacements problematizes the notion of hospitality as a humanitarian strategy as well as a moral value in refugee reception (Brun, 2010; Rozakou, 2012; Al-Abed, 2014; Carpi, 2016). Whether the ostentation of the mayor's campaign is intended or not, its success is clear in extending the state discourse to the town and making the "Middle Eastern constructions of duty-based obligations to the guest" (Chatty, 2017) visible to Western eyes. With religious overtones, the state discourse of hospitality helps to mobilize local communities and charity groups into welcoming refugees, which may often match the international humanitarian organizations in providing aid. In Kilis, a corollary to this development is the proliferation of faith-based charity groups growing into institutionalized humanitarian associations (Deniz et al., 2016).

Embodied in the assertive leadership of the Turkish President Erdoğan, this discourse draws on Islamic history and the morals that lay out hospitality as a disinterested act of offering help and protection to the person-in-need. Nevertheless, the hospitality discourse turned into humanitarian intervention converts this welcoming act into a 'policed pact' (Derrida, 2011) between the Turkish government and Syrian refugees. As a speech to the Syrian refugees in Öncüpınar camp by President Erdoğan, then Prime Minister, exemplifies, this discourse portrays the refugees as brothers in the care of the Turkish people (NTV, 2012). Heralding an anticipated victory against the Assad regime, Erdoğan not only constructs a self-image of a generous father—for the Syrian as well as Turkish nationals—but he also assumes the role of guardian protecting "suffering Muslims" as well as other religious communities in the name of Islam and Islamic brotherhood.

As illustrated by the mayor's campaign, the local state in Kilis successfully combines the domains of "institutional and everyday working of hospitality" (Dağtaş, 2017) to accommodate large numbers of refugees at the border zone. Carpi and Şenoğuz frame refugee hospitality in the Turkish and Lebanese border regions as a multi-purposed discursive strategy, which seeks to establish assertive politics of sovereignty over the social encounters between the locals and refugees (Carpi and Şenoğuz, 2018). It governs the encounters as a host–guest relationship, which serves to restore the host power over the guest. This assists in turn the Turkish government in establishing socio-spatial control over the refugees, who are compelled to accept the tacit agreement of guesthood and recognize its boundaries. Any breach to the agreement by the refugees may result in quarrels with the locals and end in the removal of refugees, as illustrated by the eviction of a group of refugees from a town neighborhood when the conflict was set out between a local and refugee family and sporadic attacks from the locals extended to adjacent refugee homes (Aljazeera, 2014).

Many locals in Kilis regarded the refugee home-making practices as a transgression of guesthood boundaries (Yaşar, 2014). They considered these practices as attesting to the prolonged stay of refugees and exacerbation of their own living conditions. They scapegoated their guests for the decrease in salaries and increase in house rents. A study reports that the question of affording housing rents, which tripled in price,[6] appears as the most challenging concern among the refugees in the town (MMP, 2017). Moreover, the locals resented the humanitarian aid offered to the refugees, while they did not receive help as much as their guests, even though they were facing economic difficulty. These feelings of grievances occasionally inciting anti-Syrian riots and racialized violence in other Turkish cities,[7] remained curbed in Kilis, as they were translated into sentiments of containment and abandonment, the locals being helplessly seized by a deeply entrenched border experience.

Humanitarian–military nexus at the border

The size of the refugee movement at the Turkish–Syrian border, difficult for local authorities and civic initiatives to cope with, opened room for the humanitarian intervention of international as well as national organizations. As Turkey shared a long borderline with Syria, Kilis being one of the most important to the northern parts of the country, crisis management involving a diverse array of actors deployed this border into a logistics rear base and buffer zone for the procurement of aid operations and refugee protection. A "humanitarian border," as coined by Walters (2011) to point to the specific relocation of humanitarian governance on the actual borders and gateways, emerges in this zone as entangled with Turkish geopolitical interests.

As Walter reminds us, the often-chaotic movements of refugees, as well as the structuring of humanitarian assistance, turns the border into a fluctuating geography. As mentioned, Kilis has become a major route for refugees, admitted at the Öncüpınar-Bab Al-Salam gate but also crossing along the well-trodden paths of minefields. Following him, the repeated displacements and misery lead to a movement out of conflict-stricken areas and make the crossing of the border a matter of life and death for refugees, though the shifting frontlines of conflict may help with voluntary returns to 'liberated' zones. The humanitarian efforts are not restricted to the camp and non-camp refugees in Kilis, while the paramount scale-up of the war causes the displacement of 7.6 million Syrians, stranded inside the country in hunger and despair in 2014 (UNHCR, 2015). As misery looms beyond the borderline, the humanitarian agencies are pressed by the moral obligation to deliver assistance to the persons in need living in the far interiors of the country and difficult to reach. The hurdles of security and unpredictability of the war situation, with a multitude of adversaries involving various proxies and agendas, shape a cross-border humanitarianism operationalizing the official and unofficial crossing points for delivery.

The United Nations (UN)-led cross-border operations in cooperation with the EU regional plan adopt a holistic focus aiming at procuring assistance to the 'Whole of Syria,' mainly from Turkey and Lebanon with Hatay (Bab Al-Hawa) and Kilis (Bab Al-Salam) as the crossing gates (OCHA, 2014). A number of international NGOs have also set up their operations at the border regions of neighboring countries to send their consignments to Syria. The Turkish–Syrian border performs its critical mission to facilitate access mainly to the northwestern opposition-controlled parts of the country as well as Aleppo's outer neighborhoods and rural Damascus, which survive prolonged sieges and short intervals of ceasefire by the Syrian government. Until the course of war and the shelling of civilian areas by the Syrian government constrained these operations, the latter were conducted under the UN Security Council resolutions that sanctioned the cross-border aid delivery with prior notification of the Syrian government and upon its consent.[8] With the UN, abandoning its position vis-a-vis the Syrian government, an imperative for seeking its consent is no longer required and the UN Office for the Coordination of Humanitarian Affairs (OCHA) deploys its office in the border city of Gaziantep, adjacent to Kilis (Ferris and Kirişçi, 2015) and coordinates with international and Syrian NGOs. However, the cross-border operations increasingly depend on the Syrian organizations working inside the country in monitoring the shipment and distribution of aid, and the aid workers, kidnapped or hit by stray bullets or deliberate attacks, remain at the mercy of fighting adversaries.[9]

The government of Turkey on the other hand introduces its individual country response to the humanitarian consequences of the war. The Turkish Red Crescent plays a prominent role in collecting aid from a diverse array of donors and programs at the national and international scale, as well as monitoring its delivery, while AFAD approves the consignments for their humanitarian nature.[10] Banning the crossing of Turkish citizens and Turkish-licensed cars to Syria in 2012, Turkey has developed a "zero point border delivery system," which consists of transporting food, in-kind aid, and medical supplies to the border gate where they are unloaded and picked up by Syrian trucks. This system is used by OCHA, international NGOs, some registered after years of deadlock applications[11] and the Istanbul-based Syrian Interim Government of National Coalition extending the opposition networks and relief activities to Syria with the help of local councils operating inside the country (Hamdan, 2017).

Turkey also extends its humanitarian relief operations to the makeshift camps of internally displaced Syrians near the border. Since the second half of 2012, for example, refugee groups waiting to enter amassed at Bab Al-Salam gate, facing Kilis. The Turkish government has sought a strategy to cater to the needs of these camps by providing tents and other supplies, rather than processing fast-track admissions for the camp dwellers, which suggests a contentious "open-border policy" above all. Soon, these camps became semi-permanent fixtures at the border, together with the widening of large-scale humanitarian projects inside Syria (Dinçer et al., 2013). Since then, humanitarian assistance to the internally displaced involves infrastructure improvement (water, sanitation, and electricity), health, educational, and emergency needs and supports the makeshift camps into a more organized structure, alongside the informally organized camps out of reach (REACH, 2014). The cross-border delivery seems to be an expansion of Turkish authority across the border zone inside Syria with the Turkish humanitarian organizations acting as proxy. The Turkish Red Crescent and the Foundation for Human Rights and Freedoms and Humanitarian Relief (IHH), a Turkish NGO with pre-dating relief experience in many countries undertake the assistance and relief operations to the camps of internally displaced Syrians. Classified as an independent NGO with an international profile, IHH has extensive "connections to the Turkish political elite" (Bilefsky and Arsu, 2010) and a wide outreach inside Syria.[12]

The camps stretching along the border between Azaz/Öncüpınar (Kilis) and Jerablus/Kargamış (Gaziantep), lying to the west of the Euphrates river, emerge as part of a humanitarian–military nexus intended by the Turkish government to secure a *de facto* "safe zone" inside Syria. However much strongly advocated by the Turkish government before its international partners, the idea of a "safe zone" did not gather sufficient support. Nevertheless, Turkey continues to insinuate into the conflicts in northern Syria for territorial control. The emergent zone of refuge for the internally displaced affords the government a leverage to justify its military expansion along the border stretch between Azaz and Jerablus. For this purpose, Turkey launched its cross-border Operation Euphrates Shield between August 2016 and March 2017 and sends troops to northern Syria in support of the Free Syrian Army to eliminate its threats (Okyay, 2017).

While the objective of clearing the border area of Jerablus from the Islamic State legitimized the operation, it was seconded by Turkey's aim to stop the advance of Syrian Democratic Forces—which is led by the Democratic Union Party (PYD) claiming an affiliation with the political platform of Abdullah Öcalan, leader of PKK—to the west of Euphrates and the opening of a northern corridor between Kurdish cantons. Turkey is concerned about a Kurdish entity bordering its territory and fears repercussions within its borders, because it might encourage the Kurdish demands for political autonomy. The self-governing region of Afrin on the westerly side of the border stretch is targeted next—a military operation with air strikes and ground troops crossing from Kilis has been launched as a remedy to the tangled interests of state powers party to the war (İdiz, 2018).

Walters argues that the securitization of the border against 'illegal entries' is inherent to the deployment of humanitarian enterprise at border zones (Walters, 2011). This sort of policing of the border is increasingly interlocked with the emerging digital technologies like biometrics. New security measures like biometric IDs were first launched in Kilis camps,[13] which would be further followed by enhanced border surveillance: a modular wall integrating a variety of high-tech security systems. A visit to the governor's office in Kilis by Deutsche Welle correspondents in March 2015 informs about the "Syrian border physical security system," a large-scale project that would ultimately fence off the Turkish–Syrian border in large part by 2018, which has started to be erected in Kilis.[14] It constitutes a separation wall, with 3.5 m height, fortified with barbed wire, modern air and ground surveillance, spotlights, watchtowers and ditches, and patrol roads.

As the examples stretching along US–Mexico, Israel–Palestine, Morocco–Spain, and Turkey–Greece borders, to enumerate a few, show, great walls have emerged as a global phenomenon. Induced by US intervention and Turkish concerns about the new geopolitics emerging within Syria, the wall is expected to spur a watertight separation between the Turkish territory and other geographies of control. Nevertheless, as Vallet and David suggest, the walls emerge paradoxically rooted in the waning of state sovereignty and the disappearance of physical borders (Vallet and David, 2012). They are part of bordering processes in the globalizing international system. Against this background, it is not easy to straightforwardly assume that the wall introduced at the Turkish–Syrian border will be impregnable. According to the geographer Leïla Vignal, the Syrian conflict renders segments of the Turkish–Syrian border into spatial envelopes of competing legitimacies among multiple actors, so long as the latter attempt to establish control over these border segments in order to sustain their warfare (Vignal, 2017). Hence, the wall's ultimate function is not a barrier between the two sides but a reterritorialization effect in the midst of territorial disintegration in Syria.

If the emergence of a humanitarian–military nexus at Kilis border causes the intensification of transnational interactions and circulations across the zone, the support of the Turkish government to the opposition groups renders the border more permeable. The allegations about sheltering of fighters in border camps, delivery of arms to the fighters by the government, and border cities-turned-outposts of logistics and financial operations for armed groups elicit reactions in domestic politics (ICGG, 2013). For instance, Öncüpınar (Kilis) as well as Apaydın (Hatay) camps have been the subject of allegations by opposition parliamentarians as sanctuaries for Syrian fighters.[15] Furthermore, despite the prevailing securitization discourse, recent years have witnessed security breaches and deadly attacks against civilians caused by 'lack of transparency' and 'insufficient monitoring' of Turkish authorities.[16] Unsurprisingly, the securitization at Kilis border has not forestalled the crossings of Islamic State militants in and out of Turkey,[17] nor the war spillovers into the town. The latter has been perpetually hit by IS rockets during the first half of 2016, with the killing of 20 people and the setting off among Kilis locals of a tendency to move next to their kin in the neighboring border city of Gaziantep.[18]

Labor markets, war economy and new trade geographies

The waging of the conflict between Syrian state and armed opposition groups in the northern parts of the country, the battle of Aleppo and air bombings of Russia and Coalition Powers led by the US military as a backup to the fighting parties respectively escalated violent clashes and led to mass displacement within Syria. As this book laid out, Kilis had strong ties of commercial and familial exchanges with Syrian Aleppo prior to the war. In addition, the rapprochement between the two countries had boosted the regional cooperation between Aleppo and the Turkish city of Gaziantep, an export giant in the region as well as a shopping and tourist destination, and highlighted Kilis as a salient transit point on the cross-border route.

The trade liberalization of the 2000s under the rule of successor son Bashar Al-Assad favored the economic position of the city of Aleppo and located it among the world networks of production, particularly in textiles (Balanche, 2014). On the other hand, the agricultural sector received the worst blow in this decade and the dissolution of agrarian production caused internal displacement within Syria and made Aleppo, with a favorable urban milieu of private investments, the preferred destination for the rural hinterland. Hence, Kilis as a familiar destination and at close proximity, is a draw for Syrians, who, initially seeking refuge in their hometown and paternal villages in northern Syria, were displaced anew because of the invasive clashes among war adversaries. Kilis offered sanctuary to many refugees who wished to keep their land and other possessions across the border within reach.

With the Syrian conflict, the business opportunities of Aleppo mainly favor the Turkish border cities and their Mediterranean hinterland as well as the metropolitan city of Istanbul. The Syrian entrepreneurs would rather move their manufacture to the nearby cities of Kilis, Gaziantep, and Mersin, a southern port city with a free trade zone, which rejoices in its good old days after the Syrian conflict. In the city of Gaziantep, for instance, a favorable business environment and an old business partnership with Turkish nationals facilitates the economic emplacement of Syrian entrepreneurs.[19] In Gaziantep and Kilis, the Syrian entrepreneurs smoothly establish their businesses whipped up by the needs of the war economy in Syria and create employment for Syrian refugees, often drawing on their former ties.[20] Though not in sheer volume, Kilis town sees the establishment of new firms with a Syrian shareholder reaching 35% of the total new firms erected in 2015—an indication that the refugee arrivals might have reanimated local businesses in Kilis, making profits from its new Syrian customers. Even though the export rates for local businesses remains a controversy, sources also point to the persistent flow of export goods to Syria, which is significantly sustained by the cross-border dispatching of humanitarian goods as exports (Özpınar et al., 2015; Öztürkler and Göksel, 2015).

In 2016, The Turkish government has issued a regulation on work permits for Syrian refugees, which nevertheless stipulates the capping of Syrian employment at 10% of the total domestic workforce. The bureaucratic and economic setbacks constrain the firms to apply work permits for refugee employees, especially in a business environment that remains largely informal (Baban et al., 2016; Kıvılcım, 2016). Whether the refugee arrivals cause an increase in the unemployment rates remains a controversial issue. Yet, the 2015 Employment Statistics indicate an increase in the unemployment rates in the statistical subregion of Gaziantep, Adıyaman, and Kilis following the EU classification of Nomenclature of territorial units for statistics, though these data are not correlated with the refugees' incorporation into the workforce. According to a study, Kilis locals who lost their jobs after the Syrian conflict blame the refugees for the loss (Öztürkler and Göksel, 2015). While the lack of individual unemployment statistics for the cities do not allow the verification of local perceptions, it could be argued that the refugee arrivals result in the displacement of low-skilled and unskilled labor, including women, especially in the informal sectors of the southeast (Del Carpio and Wagner, 2015; Kavak, 2016; Kaymaz and Kadkoy, 2016).

The locals believed that the Syrian refugees find employment in informal sectors (Paksoy et al., 2015). The refugees are paid half the average salary paid to Turkish nationals in Kilis (AI, 2014). They mainly take low-skilled or unskilled jobs in construction, service, shopkeeping, manufacturing, and agricultural sectors (Yaşar, 2014). While an increasing number of firms absorb the increased workforce due to the incoming of refugees, this does not parallel the levels of rise in income levels. In a comparative perspective, the lower share of gross domestic product (GDP) per capita in Kilis puts the local resources allocated for refugees under constraints and the town is among the cities with the lowest shares of refugee income. In Kilis, the monthly household income is only 216 USD (641 TL), while the refugees in Gaziantep obtain three times higher on average (Balcılar and Nugent, 2018). The economic constraints also lead Syrian families to let their children drop out of school in order to participate in the labor force, thereby exacerbating child labor in the town (AI, 2014).

The Syrian conflict engenders an "asymmetrical border regime" (Montabone, 2016) connecting the two sides of the border with territorial disintegration and military strife in the south. The new bordering practices, as the population upsurge in the town made clear, favors an illicit movement from south to north, reversing the former practices of cross-border laboring and petty-trade. While the landowning notables in Kilis used to exploit the border peasants as agricultural laborers in their properties at the Syrian side of the border until the 1960s, the new border regime introduced after the Syrian conflict allows the daily crossings of Syrians to work as farmhands for the smallholders in Kilis. The porous border encourages the Syrian refugees to engage in petty trade and sell their goods in the town streets. My visit to one of the former shopping passages, left idle for years, also revealed the crowding of Syrian shops, all full of most-traded staple goods of cigarettes and tea as well as other items.

My visit to the town ran into a time after the Turkish government closed its border points with Syria on March 8, 2015 (EU, 2016). Expecting a sharp decrease in the number of border crossings, I observed a maze of paths trotted by the Syrian refugee groups along the border villages in close proximity to the minefields the same year during the mid-summer Ramadan holiday. As I was passing by a local transport company operating between Hatay and Gaziantep, two cities neighboring Kilis as well as the border, the refugee movement across the border returning from visiting their family members in Syria enhanced the erratic nature of these journeys with delayed departures and unexpected stops along the way. The driver, a suntanned man in his thirties, typically sought to pick up more customers on his way. They usually do in the small Turkish cities, where the number of commuting passengers are never enough to fill the vehicles. Surprisingly however, this time the driver aimed at profiteering from the refugees as customers.

The irregular crossings mounting during the religious holiday, illustrated how this unregulated transportation was coveted among local transport companies as well as border inhabitants as extra earnings. My driver was indeed a primary school teacher who started to work in his uncle's company during the summertime in order to make the most of commuting refugees. Tractors, vans, and cars transported passengers from the minefields to Kilis center or the parting of the Gaziantep road where the passengers would be picked up by other vehicles. One could read the news about the increase in local taxi companies, formerly conveying passengers across the border, now dropping off camp and non-camp refugees between locations of their choice (Kilis Postası, 2014). This lucrative business also pulled in some nation-wide transport companies to open offices in the countryside. Along the way crossing the Kurd Dagh, the villages near the road hosted these companies as well as groups of refugees, waiting to leave.

My journey helped to detect that the previous smuggling networks rapidly gave way to politically recognized networks of migration. The latter lacks the characteristics of smuggling as an ethically and locally-bound practice and turns the smuggling of Syrian refugees across the border into a capacity that any border actor with enough power can take hold and solicit a deal without seeking the consent of the opposite party of the deal. The refugees use unofficial crossings which are controlled by various opposition groups in northern Syria. Control of the crossing points provides political and material advantages to these groups in their warfare. (Vignal, 2017). On the other hand, the refugee movement across the border fosters a war economy that operates through complicated border arrangements among the opposition, smugglers at each side of the border, the Turkish military patrol and the refugees. The interviews with border peasants suggest that the opposition groups broker the agreement with the Turkish officers and the smugglers help the refugees crossing, while each party held a different share from the money paid by the refugees. The crossing points mentioned by the interviewees not only refer to the former trade routes but they also attest to the reports on the unofficial gates along the Syrian border where the coalitions of opposition and armed groups secure their access. In mid-2015, the refugees were channeled across three crossing points, namely Kurd Dagh, Elbeyli, and Arpakesmez-Akıncı border segment in the middle. YPG (*Yekîneyên Parastina Gel*, People's Protection Units), which is the leading component of the opposition Syrian Democratic Forces (SDF), Al Nusra—later joining the Islamic State—and the Northern Storm Brigade (*Liwa Asifat al-Shamal*) secured access to the border.[21] The competition for control shifts the frontlines ever since and leads to the forming of new alliances.

The Syrian conflict shapes new trade geographies along the Turkish–Syrian border. The Turkish borderlands turn into an outpost for these groups where they can take refuge and grow shadow trade. For example, the drug traffic, mainly Captagon—an easy and inexpensive amphetamine to produce—is believed to have circulated across the Turkish–Syrian border in order to supply the finances of armed groups, until the Turkish government increased counter-operations between 2014 and 2015 (Kravitz and Nichols, 2016). Syria, which has already been a hub of production for Captagon before the war became geared towards transnational organized drug production. The drug trafficking has been channeled across the Turkish–Syrian border for shipment to the Gulf States via Mersin port. Although the cross-border traffic is being intercepted by the Turkish authorities that are forced to fence off the border against 'terrorists' in northern Syria, it is not plausible to infer its complete eradication.

While the Syrian conflict has dramatically blocked the land routes supplying the Middle Eastern markets, the war has enabled new maritime routes circumventing Syria (Montabone, 2016). The goods delivered to the Turkish ports of Mersin and İskenderun (Hatay) are repacked for shipment to the Israeli port of Haifa, where the goods land again for transportation to the countries of the Middle East. This maritime trade, providing an alternative to the former land transport, both re-enacts the old maritime route of smuggling from Beirut port to the eastern Mediterranean shores of Turkey in the early 1980s and creates a corridor space between the southeastern Mediterranean coasts and Arabian Peninsula, promoting an 'off-the record' trade crossing through the Israeli gateways (Plonski, 2017). Along this corridor also extends the global supply chains between the West and Far East Asia—hence the reason for China's investments in the infrastructure improvement of Jordanian Aqaba port as well as national railway networks (Evron, 2016). Israel and neighboring countries of the Middle East lack former trade agreements. Therefore, the goods transit Israeli customs without a paper trail ending directly at Jordanian customs. The free trade zones and ports in the region, as former hotspots of "extra-legal forms of trade" (Moore and Parker, 2007) gain importance as gateways to the global trade corridor. Clandestine trade—i.e. oil and drug—set in motion by the warmongers is channeled across the segments of the Turkish–Syrian border towards these trade zones and ports.

While the rumours in Kilis suggest clandestine trade between each side of the border to supply the armed groups in Syria, it is not easy to discern to what extent the borderland remains connected to the clandestine trade networks. Still, however, although it may seem speculative, the new global trade corridor might conceivably alter the life prospects of the border dwellers creating new labor geographies, turning the refugees into cheap labor to feed the global supply chains. The re-enacted plans for the erection of a joint organized industrial zone specialized in textiles on the territorial boundary between Kilis and Gaziantep, where the latter seeks to take advantage of the economic incentives given to the former within the framework of "priority development areas," constitute such a potential (Hürriyet, 2016). The industrial zone will be connected by a new highway and tunnel access to the İskenderun port for shipment to further destination markets.

Concluding remarks

The Syrian conflict radically changes the political rationalities and technologies of border-making, moving it away from a Westphalian logic. Against the background of sub- and supra-national border dynamics, the new international relations compel

Turkey to secure its border, while furthering at the same time the segmentation of border zones, allowing various opposition groups to take leverage from border access. The Turkish–Syrian borderlands are salient to military warfare, global humanitarian relief, and refugee movements, which do not necessarily undermine the Turkish sovereignty but rather give its migration policies and geopolitical interests certain momentum. The conflict also shifts old historical trade routes into global corridors and creates new infrastructures, extending global supply chains across Middle Eastern geography in a state of flux. This chapter aimed to shed light on these developments through the prism of Kilis border.

Prolonged discussions about refugee deals, humanitarianism, and regional peace with a focus on border zones tend to cast a cloud over the actual lives of border dwellers. Although the post-conflict developments pinpoint the town of Kilis on a transnational space, their impact among the inhabitants, both Turkish and Syrian nationals, do not gather matching attention. My fieldwork in the town during the period January 2011–June 2012 disclosed beneath the secretive attitudes of town dwellers, stigmatized as smugglers, not only their concerns about maintaining their status, protecting their economic interests, and eluding law enforcement, but also about building a strong sense of community on the basis of their common experience of living at the border. Moreover, their sense of attachment straddled the border and encompassed dwellers of the Syrian side, who they referred to as kin whether or not related by blood. In this respect, the field experience culminated in the idea that the border is not only a barrier that distinguishes border communities along ethnic and religious lines, but is an opportunity for earning livelihood, realizing aspirations, and establishing bonds of trust, reciprocity, and affinity.

The encounters between Kilis locals and Syrian refugees, framed as a host–guest relationship, undermine these bonds, while humanitarian aid as well as labor exploitation by informal businesses pitting the two groups against each other, compromise the sense of community and familiarity. Grey areas appearing between survival marriages and sex trafficking, compassion to the refugees and their abuse, alliance and partnership are an outcome of the reshuffling of values, practices, and relationships inflicted by the new forms of mobility and containment at the border. Although these forms foster opportunities and mend ties between the locals and refugees, they also turn them into separate enclaves at the border whose co-habitation becomes ever more difficult to accommodate.

Notes

1   The population of Turkish nationals in the address-based registration system for 2016 is 130,655 with more than 100,000 registered in the central district. Yet, it is worth reminding that since Kilis is a transit location, together with a continual margin of unregistered refugees and persistent decline in the camp population, the number of Syrian refugees remains variable. In December 2017, the number of camp refugees, especially those registered in the rural Elbeyli, has significantly dropped—between 2014 and 2017, it decreased from 24,062 to 14,280 in Elbeyli and from 37,477 to 26,395 in both camps (AFAD, 2017). It is likely that the camp refugees opt to leave the Elbeyli camp and self-settle in Kilis as well as other cities for better employment opportunities.

2 Erdoğan suggests that this perception does not reflect the reality, referring to a statement by the Kilis Governorship (Erdoğan, 2014, p. 70). The city governor shared his study on hospitals in which there is a 3% share of Syrians in health services. According to him, the Syrians in the emergency room are taken by the locals as if they were in all the departments of the hospitals.

3 Kilis'teki Suriye: Sorunların Tespit ve Çözümlerine İlişkin Rapor, Kilis Ortak Akıl Topluluğu (2013).

4 Cf. Özden (2013). Özden states that in the border city of Hatay where historical bonds with Syria are much stronger because of the local Arab Alevi community, the latter talked about their cultural differences with "the peasant, non-urban and uneducated Syrians" (Özden, 2013, p. 10). These differences were also played out as sectarian tensions between local Alevis and Sunni refugees. On the other hand, Deniz et al. (2016) argue that the locals and refugees sharing the same ethnic background in Kilis may develop solidarity and cooperation, undermining the social boundaries between the two communities. See also Şenoğuz (2017) for the grassroots solidarity among Turkish and Syrian Kurds during the siege of Kobanê, a northern Syrian town under the attack of Islamic State in 2014.

5 The EU–Turkey Statement, notoriously known as the Deal was enforced on March 18, 2016 to resettle the asylum seekers in Europe to Turkey declared as a "safe third country" in exchange for Syrian refugees to be relocated in European countries. See Heck and Hess (2017) and Soykan (2016) for controversies over the deal.

6 A 2013 USAK (Uluslararası Stratejik Araştırmalar Kurumu) and Brookings Institute report states that the housing rents ranging between 200 and 300 TL quickly rose between 700 and 1000 in Kilis (Dinçer et al., 2013).

7 Şenoğuz (2017); Şimşek (2015); Özden (2013).

8 See the UN Security Council Resolution 2165/2191/2258/2332 in OCHA (2017).

9 The hospitals supported by Médecins Sans Frontiers (MSF) in Azaz and Idlib are repeatedly attacked by several opposition groups as well as the Syrian government, a deep concern that forced MSF to stop sharing with the latter the location of its facilities (Shaheen, 2016).

10 Interview with AFAD head of education unit in Gaziantep, May 18, 2016.

11 I observed, for instance, that MSF needed to keep a low-profile at the border when I took a short-term translation job to its staff in Kilis for the preparations of a hospital inside Syria in late 2012. MSF, as well as several other international NGOs, were registered later on (IRIN News, 2015).

12 The IHH Hatay coordinator boldly claims that the charity organization has access to 18% of Syria (ICGG, 2013).

13 See Kirişçi and Ferris. In 2015, the Turkish authorities extended the use of biometric information to non-camp dwellers registered with DGMM (General Directory of Migration Management) (World Bank, 2015).

14 According to Okyay (2017), the wall reached 383 km by early 2017.

15 Mahmut Toğrul from the Peoples' Democracy Party, an opposition group in the parliament, raised suspicions about a camp resident in military uniform and in constant conversation on his three cell phones while he visited the Öncüpınar border crossing in 2015 (Taştekin, 2016). For mounting criticisms about Apaydın camp, see the MERIP (Middle East Research and Information Project) Report (Ilgıt and Davis, 2013).

16 Toğral-Koca (2015) illustrates the practices of Turkish authorities by reporting about the pushback operations, violence against refugees as well as the lack of adequate screening of these violations at the border.

17 See again Taştekin (2016) for Islamic State militants sweeping through Kilis border.

18 Kilis town received about 70 rocket strikes (Girit, 2016). The migration statistics indicate an out and return migration among the locals, whereas the data about the refugee movement are not available.

19 According to the figures reported in 2016 by the Chamber of Trade, about 614 Syrian businesses established after the conflict accommodated nearly 350,000 refugees (Hürriyet Daily News, 2016).
20 A young Syrian entrepreneur, owner of a garment workshop in Gaziantep's Ünaldı, a deregulated industrial site in the close district of the urban center infamously known for informal and precarious employment, explained that he could manage to gather 70% of his former employees after the war, which used to work in his factory in Aleppo.
21 See Al-Tamimi (2015) for the inter-rival competition to secure access to the gates in this border segment.

References

AFAD (2017) Geçici Barınma Merkezi Raporları, 11.12.2017. Retrieved 22 December 2017 from www.afad.gov.tr/upload/Node/2374/files/11_12_2017_Suriye_GBM_Bilgi_Notu.pdf.

AI (Amnesty International) (2014) *Struggling to Survive Refugees from Syria In Turkey*, London: Amnesty International.

Al-Abed, O. (2014) "The Discourse of Guesthood: Forced Migrants in Jordan", pp. 81–100 in A. Fábos, A. and R. Isotalo, *Managing Muslim Mobilities*, London: Palgrave MacMillan.

Aljazeera (2014) "Kilis'de Suriyeliler tahliye edildi", 15.8.2014. Retrieved 8 December 2017 from www.aljazeera.com.tr/haber/kiliste-de-suriyeliler-tahliye-edildi.

Al-Tamimi, A. J. (2015), Special Report: Northern Storm and the Situation in Azaz (Syria), *Middle East Review of International Affairs (MERIA Journal)*. Retrieved 25 December 2017 from www.aymennjawad.org/15865/special-report-northern-storm-and-the-situation.

Ay, M. (2014) *The Mass Influx of Syrian Refugees to Turkey*, Emergency and Disaster Reports, University of Oviedo, 1(2): 2–53.

Baban, F., Ilcan, Z. and Rygiel, K. (2016) "Syrian Refugees in Turkey: Pathways to Precarity, Differential Inclusion, and Negotiated Citizenship Rights", *Journal of Ethnic and Migration Studies*, 43(1): 41–57.

Balanche, F. (2014) "Alep et ses territoires: une métropole syrienne dans la mondialisation", pp. 39–65 in J. David and T. Boissière, (eds.), *Alep et Ses Territoires: Fabrique et politique d'une ville (1868-2011)*, Presse de l'Ifpo.

Balcılar, M. and Nugent, J. B. (2018) *The Migration of Fear: An Analysis of Migration Choices of Syrian Refugees*, Discussion Paper 15–36, Fagamusta, Eastern Mediterranean University, North Cyprus.

Biehl, K. et al. (2016) *Needs assessment Report: Technical Assistance for a comprehensive needs assessment of short and medium to long term actions as basis for an enhanced EU support to Turkey on the refugee crisis*, Service Contract No. 2015/366838, EUROPEAID/129783/C/SER/multi, European Commission. Retrieved 24 January 2018 from www.avrupa.info.tr/fileadmin/Content/2016__April/160804_NA_report__FINAL_VERSION.pdf.

Bilefsky, D. and Arsu, Ş. (2010) "Sponsor of Flotilla Tied to Elite of Turkey", *The New York Times*, 15.7.2010. Retrieved 25 January 2018 from www.nytimes.com/2010/07/16/world/middleeast/16turkey.html?_r=1.

Brun, C. (2010) "Hospitality: Becoming 'IDPs' and 'Hosts' in Protracted Displacement", *Journal of Refugee Studies*, 23(3): 337–355.

Cagaptay, S. (2014) *The Impact of Syrian Refugees on Southern Turkey*, Policy Focus 130, The Washington Institute for Near East Policy, Washington, pp. 1–32.

Carpi, E. (2016) "Against Ontologies of Hospitality. About Syrian Refugeehood in Northern Lebanon", Middle East Institute. Retrieved 12 December 2017 from www.mei.edu/content/map/against-ontologies-hospitality-about-syrian-refugeehood-northern-lebanon.

Carpi, E. and Şenoğuz, P. H. (2018) "Refugee Hospitality in Lebanon and Turkey: On Making the 'Other'", Special Issue on Syrian Refugees: Facing Challenges, Making Choices, *International Migration Journal*, forthcoming.

Chatty, D. (2017) "The Duty to Be Generous (Karam): Alternatives to Rights-Based Asylum in the Middle East", *Journal of the British Academy*, 5: 177–199.

CNN Türk (2016) "Kilis Nobel Barış Ödülü'ne Aday", 7.3.2013. Retrieved 24 December 2017 from www.cnnturk.com/turkiye/kilis-nobel-baris-odulune-aday.

Dağtaş, S. (2017) "Whose Misafirs? Negotiating Difference along the Turkish-Syrian Border", *International Journal of Middle East Studies*, 49: 661–679.

Del Carpio, X. D. and Wagner, M. (2015) *The Impact of Syrian Refugees on the Turkish Labor Market*, Policy Research Working Paper 7402, World Bank.

Deniz, Ç. A., Ekinci Y. and Hülür, B. (2016) *"Bizim Müstakbel Hep Harap Oldu": Suriyeli sığınmacıların gündelik hayatı, Antep-Kilis çevresi*, İstanbul: Bilgi Üniversitesi Yayınları.

Derrida, J. (2011) The Principle of Hospitality, *Parallax*, 11(1): 6–9.

Dinçer, O. B. et al. (2013) *Suriyeli Mülteciler Krizi ve Türkiye: Sonu Gelmeyen Misafirlik*, Ankara: USAK and Brookings Institute.

ECPAT (2015) *Global Monitoring: Status of Action against Commercial Sexual Exploitation of Children*, 2nd Edition, ECPAT International (End Child Prostitution, Child Pornography and Trafficking of Children for Sexual Purposes).

Erdoğan, M. (2014) "Perception of Syrians in Turkey", *Insight Turkey*, 16(4): 65–75.

Erkoç, T. (2013) "Refugee-Asylum Seeker Policy of Turkey in The Light of Recent Developments", pp. 39–44 in E. Akçay and F. Alimukhamedov (eds.), *Refugee-Asylum Seeker Policy of Turkey in the Light of Recent Developments*, Workshop Proceedings, Ankara: The Journalists and Writers Foundation Press,

EU (2016) EU *Humanitarian Implementation Plan – Syrian Regional Crisis*, Ref. Ares (2016)1865591 – 20/04/2016. Retrieved 25 January 2018 from http://ec.europa.eu/echo/files/funding/decisions/2016/HIPs/HIP%20V2%20FINAL.pdf.

Evron, Y. (2016) "Can China Participate in Middle East Stabilization Efforts by Supporting Regional Connectivity?", *Asia Pacific Bulletin*, no. 363.

Ferris, E. and Kirişçi, K. (2015) *From Turkey to Syria: The Murky World of Cross-border assistance*, The Brookings Institution. Retrieved 23 December 2017 from www.brookings.edu/opinions/from-turkey-to-syria-the-murky-world-of-cross-border-assistance/.

Girit, S. (2016) "Syria Conflict: Kilis, the Turkish Town Enduring IS Bombardment", BBC, 9 May 2016. Retrieved 14 November 2017 from www.bbc.com/news/world-europe-36245505.

Green, S. (2012) "A Sense of Border: The Story So Far", pp. 573–592 in T. M. Wilson and H. Donnan (eds.), *A Companion to Border Studies*, Chichester: Wiley-Blackwell.

Hamdan, A. (2017) "Stretched Thin: Geographies of Syria's Opposition in Exile", *Refugees and Migration Movements in the Middle East*, University of California Press, Berkeley. Retrieved 25 November 2017 from https://pomeps.org/wp-content/uploads/2017/03/POMEPS_Studies_25_Refugees_Web.pdf.

Heck, G. and Hess, S. (2017) "Tracing the Effects of the EU-Turkey Deal: The Momentum of the Multi-Layered Turkish Border Regime", *Movements: Journal for Critical Migration and Border Regime Studies*, 3(2): 35–56.

Herwig, R. (2017) "Strategies of Resistance of Syrian Female Refugees in Şanlıurfa", *Movements: Journal for Critical Migration and Border Regime Studies*, 3(2): 177–192.

Hürriyet (2016) "Gaziantep-Kilis Arasındaki OSB'ye Bakanlıktan Onay", 14.10.2016. Retrieved 14 November 2017 from www.hurriyet.com.tr/gaziantep-kilis-arasindaki-osbye-bakanliktan-o-40248929.

Hürriyet Daily News (2016) "AKP Moves to Nominate Turkish Town Hosting Syrians for Nobel Prize", 8.2.2017. Retrieved 24 January 2018 from www.hurriyetdailynews.com/akp-moves-to-nominate-turkish-town-hosting-syrians-for-nobel-prize-94882.

ICGG (2013) *Blurring the Borders: Syrian Spillover Risks for Turkey*, Report No. 225, 30 April 2013.

Ilgıt, A. and Davis, R. (2013) The Many Roles of Turkey in the Syrian Crisis, Merip, January 28, 2013. Retrieved 21 January 2018 from www.merip.org/mero/mero012813.

İdiz, S. (2018) "Erdogan scores big with Afrin operation, but problems remain", Al-Monitor, 23 January 2018. Retrieved 25 January 2018 from: www.al-monitor.com/pulse/originals/2018/01/turkey-syria-erdogan-scores-big-with-afrin-operation.html.

IRIN News (2015) "Turkish NGO Move Boosts Syria Aid Delivery", 25.6.2015. Retrieved 25 January 2018 from www.irinnews.org/news/2015/06/25/turkish-ngo-move-boosts-syria-aid-delivery.

Kavak, S. (2016) "Syrian Refugees in Seasonal Agricultural Work: A Case of Adverse Incorporation in Turkey", *New Perspectives on Turkey*, 54: 33–53.

Kaymaz, T. and Kadkoy, T. (2016) *Syrians in Turkey – The Economics of Integration*, Alsharq Forum Expert Brief.

Kıvılcım, Z. (2016) "Legal Violence against Female Syrian Refugees in Turkey", *Feminist Legal Studies*, 24(2): 193–214.

Kilis Postası (2014) Kilis'te Taksi Durakları Artıyor, 26.11.2014. Retrieved 25 January 2018 from www.kilispostasi.com/kilis-te-taksi-duraklari-artiyor/1207690/.

Kilis'teki Suriye: Sorunların Tespitine ve Çözümlerine İlişkin Rapor, Kilis Ortak Akıl Topluluğu, 2013.

Kravitz, M. and Nichols, W. (2016) "A Bitter Pill to Swallow: Connections Between Captagon, Syria, and the Gulf", *Journal of International Affairs*, 18.5.2016. Retrieved 13 September 2017 from https://jia.sipa.columbia.edu/bitter-pill-swallow-connections-captagon-syria-gulf.

Mazlum-Der (2014) *Report on Syrian Women Refugees Living Out of the Camps*, Mazlum-Der Women Studies Group. Retrieved 25 January 2018 from http://istanbul.mazlumder.org/webimage/report-of-syrian-women-refugees-living-out-of-the-camps.pdf.

McClelland, M. (2014) "How to Build a Perfect Refugee Camp", *The New York Times*, 13 February 2014. Retrieved 24 January 2018 from www.nytimes.com/2014/02/16/magazine/how-to-build-a-perfect-refugee-camp.html.

MMP [Mixed Migration Platform] (2017) *Refugee, Asylum-Seeker and Migrant Perception in Gaziantep and Kilis*, 29 June 2017. Retrieved 25 December 2017 from http://groundtruth-solutions.org/wp-content/uploads/2017/07/MMP_Turkey_R1_Gaziantep.pdf.

Montabone, B. (2016) "The Wartime Emergence of a Transnational Region Between Turkey and Syria", pp. 181–198, in L. Vignal (ed.), *The Transnational Middle East: People, Places, Borders*, Oxford, New York: Routledge.

Moore, P. and Parker, C. (2007) "The War Economy of Iraq", MERIP, Vol. 37.

NTV (2012) "Erdoğan'dan Suriyelilere: Zaferiniz uzak değil", 6.5.2012. Retrieved 17 December 2017 from www.ntv.com.tr/turkiye/erdogandan-suriyelilere-zaferiniz-uzak-degil,eh4IpDgi20OApyeeBVBIdQ.

OCHA (2014) *Humanitarian Bulletin: Syria Operations from Turkey*, Issue 4, 20 Sept – 03 Oct 2014. Retrieved 25 January 2018 from www.humanitarianresponse.info/system/files/documents/files/20141003_Humanitarian%20Bulletin.pdf.

OCHA (2017) Fact Sheet: United Nations Cross-border Operations from Turkey to Syria. Retrieved 25 January 2018 from www.humanitarianresponse.info/system/files/documents/files/20170331_fact_sheet_unscr2165-2191-2258.pdf.

Okyay, A. (2017) "Turkey's Post-2011 Approach to its Syrian Border and its Implications for Domestic Politics", *International Affairs*, 93(4): 829–846.

Orhan, O. and Gündoğar, S. S. (2015) *Effects of the Syrian Refugees on Turkey*, Orsam Report No: 195. Retrieved 25 January 2018 from www.orsam.org.tr/files/Raporlar/rapor195/195eng.pdf.

Özden, Ş. (2013) *Syrian Refugees in Turkey*, Migration Policy Center Research Report 2013/05, San Domenico di Fiesole: European University Institute.

Özpınar, E., Başıhoş, S. and Kulaksız, A. (2015) *Trade Relations with Syrian after the Refugee Influx*, TEPAV Report, No. 201527.

Öztürkler, H. and Göksel, T. (2015) *The Economic Effects of Syrian Refugees on Turkey: A Synthetic Modelling*, Orsam Report, No. 196.

Paksoy, M. et al. (2015) Suriyelilerin Ekonomik Etkisi: Kilis İli Örneği, *Birey ve Toplum*, 5(9): 143–173.

Plonski, S. (2017) The Ontology of a "Train to Nowhere", Border, Fences, Wirewalls: Assesing the Changing Relationship of Territory and Institutions, Max Planck Institute for the Study of Religious and Ethnic Diversity, Göttingen, Germany, 19–20 October.

REACH (2014) Syria crisis: Camps and informal Settlements in Northern Syria, Humanitarian baseline review, June. Retrieved 25 January 2018 from https://reliefweb.int/report/syrian-arab-republic/syria-crisis-camps-and-informal-settlements-northern-syria-humanitarian.

Rozakou, K. (2012) "The Biopolitics of Hospitality in Greece. Humanitarianism and the Management of Refugees", *American Ethnologist*, 39(3): 562–577.

Şenoğuz, P. H. (2017) "Border Contestations, Syrian Refugees and Violence in the Southeastern Margins of Turkey", *Movements: Journal for Critical Migration and Border Regime Studies*, 3(2): 163–176.

Shaheen, K. (2016) "MSF Stops Sharing Syria Hospital Locations after 'Deliberate' Attacks", *Guardian*, 18 February 2016. Retrieved 20 January 2018 from www.theguardian.com/world/2016/feb/18/msf-will-not-share-syria-gps-locations-after-deliberate-attacks.

Şimşek, D. (2015) "Anti-Syrian Racism in Turkey", *Open Democracy*, 27.1.2015. Retrieved 31 October 2017 from www.opendemocracy.net/north-africa-west-asia/dogus-simsek/antisyrian-racism-in-turkey.

Soykan, C. (2016) "Turkey as Europe's Gatekeeper – Recent Developments in the Field of Migration and Asylum and the EU-Turkey Deal of 2016", pp. 52–60 S. Hess et al. (eds.), *Der Lange Sommer der Migration: Grenzregime III*, Assoziation A.

Taştekin, F. (2016) "How the Islamic State Is Still Sweeping Through Syria-Turkey Border", *Al-Monitor*, 1.2.2016. Retrieved 28 December 2017 from www.al-monitor.com/pulse/originals/2016/02/turkey-syria-greedy-smugglers-islamic-state.html.

Toğral Koca, B. (2015) Deconstructing Turkey's "Open Door" Policy towards Refugees from Syria, *Migration Letters*, 12(3): 209–225.

UNHCR (2015) *Global Trends: Forced Displacement in 2015*. Retrieved 25 January 2018 from www.unhcr.org/576408cd7.pdf.

Vallet, E. and David, C. (2012) "Introduction: The (Re)Building of the Wall in International Relations," *Journal of Borderlands Studies*, 27(20): 111–119.

Vignal, L. (2017) "The Changing Borders and Borderlands of Syria in a Time of Conflict", *International Affairs*, 93(4): 809–827.

Walters, W. (2011) "Foucault and Frontiers: Notes on the Birth of the Humanitarian Border", pp. 138–164 in U. Bröckling, S. Krassman and T. Lemke (eds.), *Governmentality: Current Issues and Future Challenges*, London: Routledge.

World Bank (2015) *Turkey's Response to the Syrian Refugee Crisis and the Road Ahead*, No. 102184. Retrieved 25 January 2018 from http://documents.worldbank.org/curated/en/583841468185391586/pdf/102184-WP-P151079-Box394822B-PUBLIC-FINAL-TurkeysResponseToSyrianRefugees-eng-12-17-15.pdf.

Yaşar, R. M. (2014) *Kilis'te Sığınmacı Algısı: Toplumsal Otizm ve Ötekileştirme Sürecinin İlk Görünümleri*, Kilis: Kilis 7 Aralık Üniversitesi Matbaası.

Appendix
Interviewee profiles and explanatory notes for oral history and in-depth interviews (January 2011–June 2012)

#	Substitute Name	Gender	Status/Occupation	Birth Year Intervals	Location of Interview	Date of Interview
I. Traditional notables						
1	Yüksel	female	trade notable, housewife	1925–1930	town center	22.2.2011
2	Ayhan	female	traditional landed notable, housewife	1920–1925	town center	5.11.2011
3	Hikmet	female	traditional landed notable, housewife	1925–1930	town center	25.3.2011
4	Rauf	male	traditional landed notable, pharmacist	1950–1955	town center	14.4.2011
5	Ferit	male	traditional landed notable, dentist	1935–1940	Istanbul	22.4.2011
6	Şükrü	male	trade notable, farmer	1940–1945	town center	17.1.2011
7	Latife	female	traditional landed notable, housewife	Born in 1911	Istanbul	20.7.2011
8	Beyhan	female	trade notable, housewife	1925–1930	town center	15.4.2011
9	Vefa	male	trade notable, retiree	1940–1945	town center	13.5.2011
10	Asuman	female	trade notable, housewife	1920–1925	town center	20.5.2011
11	Gencay	female	trade notable, housewife	1925–1930	town center	9.3.2011
12	Behire & Tijen (sisters)	female	trade notable, passage owner	1935–1940	town center	23.5.2011 and 4.6.2011
12		female	trade notable, passage owner	1930–1935	town center	
13	Mahmut	male	traditional landed notable, international taxi company owner	1975–1980	town center	21.10.2011 and 4.2.2012
14	Faik	male	traditional landed notable, retiree	1950–1955	town center	27.4.2011

(*Continued*)

#	Substitute Name	Gender	Status/Occupation	Birth Year Intervals	Location of Interview	Date of Interview
15	Nihal	female	traditional landed notable, veterinary	1945–1950	town center	2.4.2011
16	Bahadır	male	traditional landed notable, farmer	1950–1955	town center	26.2.2012

II. New wealth and the middle class

#	Substitute Name	Gender	Status/Occupation	Birth Year Intervals	Location of Interview	Date of Interview
17	Zeynep	female	rural family background, housewife	1925–1930	Gaziantep	7.3.2012
18	Mustafa (Koyuncu family)	male	rural family background, retiree	1935–1940	town center	19.5.2011 and 23.5.2011
19	İsmail (Koyuncu family)	male	rural family background, retiree	1930–1935	town center	26.11.2011
20	Enver	male	rural family background, owner of gas station	1935–1940	town center	17.6.2011
21	Alaattin	male	rural family background, shop keeper	1945–1950	town center	25.7.2011
22	Yasin	male	middle class family of mercantile origin, owner of a white goods store and wholesale shop for Syrian passengers	1960–1965	town center	15.11.2011
23	Vahab	male	rural family background, chieftain of a Kurdish tribe of Syrian origin and former deputy and mayor	1930–1935	town center	14.4.2011
24	Ata	male	rural family background, muhktar	1950–1955	town center	4.12.2011
25	Murtaza	male	rural family background, cross-border transport company owner	1975–1980	town center	25.1.2012
26	Urup İsmail	male	rural family background, sheikh of smugglers' sharia	1945–1950	town center	7.1.2012
27	Cemil	male	middle class family, lawyer and former senator	1940–1945	Ankara	6.6.2011

#	Substitute Name	Gender	Status/Occupation	Birth Year Intervals	Location of Interview	Date of Interview
28	Bahri	male	middle class family, lawyer	1970–1975	town center	2.11.2011
29	Ekrem	male	middle class family, local historian	1930–1935	town center	21.5.2011
30	Celal	male	middle class family, local historian	1970–1975	town center	16.5.2011
31	Cemal (interviewed with Mevlüt)	male	middle class family, retired teacher	1950–1955	town center	27.07.2011 and 21.10.2011

III. Urban and rural poor

#	Substitute Name	Gender	Status/Occupation	Birth Year Intervals	Location of Interview	Date of Interview
32	Serdar	male	subcontract employee at public hospital and small-scale trader	1980–1985	town center	26.08.2011
33	Asiye	female	small-scale trader	1930–1935	town center	29.08.2011 and 26.12.2011
34	Leyla	female	small-scale trader	1965–1970	town center	11.3.2012
35	Hamit	male	grocer and small-scale trader	1980–1985	town center	16.1.2012
36	Mevlüt	male	retiree	1930–1935	town center	27.07.2011 and 21.10.2011
37	Talip	male	small-scale trader	1985–1990	town center	14.12.2011
38	Müşir	male	grocer	1970–1975	town center	13.2.2012
39	Rıdvan	male	buffet owner	1970–1975	town center	occasional visits during September 2011–February 2012
40	Bahtiyar	male	small-scale trader	1975–1980	town center	19.1.2011
41	Hayrullah & Şükriye (husband and wife)	male	border villager	1940–1945	border village	13.3.2012
		female	border villager	1945–1950	border village	
42	İlyas	male	border villager	1940–1945	border village	20.3.2012
43	Nesim	male	border villager and small-scale trader	1975–1980	border village	15.3.2012
44	İhsan	male	retiree worker	1925–1930	town center	12.5.2011

IV. Governmental and semi-public institutions (conducted during April–May 2012)

45	Kilis Chamber of Industry and Trade (individual interviews with head of chamber and general secretary)

(*Continued*)

#	Substitute Name	Gender	Status/Occupation	Birth Year Intervals	Location of Interview	Date of Interview
46	Social Security Institution Provincial Directorate					
47	Turkish Employment Agency Provincial Directorate					
48	Provincial Directorate of Agriculture					
49	Provincial Directorate of Social Assistance and Solidarity					
50	Deputy Governor (responsible from Customs)					
51	City Municipality (Mayor)					
52	Chief at Provincial Customs Directorate					

V. Informal dialogues

53	11th grade students at vocational high school. Discussions during introductory conversations at the beginning of and throughout the Fall semester 2011.
54	Occasional gatherings with women: tea party, indoor women's day organization, and other social events by two women NGOs, home gatherings, etc.
55	Group of peasants in border villages of Kurd Dagh: Though I was accompanied by a local man who assisted me in introducing myself by speaking to them in Kurdish, they refrained from talking. I collected narratives of village history and difficulties of being a border dweller. The visits included four villages in the early weeks of August 2011.

Index

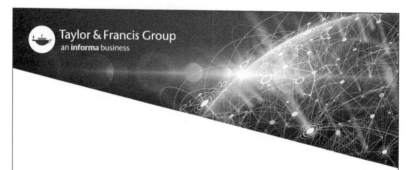